WORKPLACE HEALTH AND SAFETY CRIMES

Bill C-45 and the New Westray Criminal Offences

Norm Keith

LexisNexis
Butterworths

Workplace Health and Safety Crimes
© LexisNexis Canada Inc. 2004
April 2004

Members of the LexisNexis Group worldwide

Canada	LexisNexis Canada Inc, 75 Clegg Road, MARKHAM, Ontario
Argentina	Abeledo Perrot, Jurisprudencia Argentina and Depalma, BUENOS AIRES
Australia	Butterworths, a Division of Reed International Books Australia Pty Ltd, CHATSWOOD, New South Wales
Austria	ARD Betriebsdienst and Verlag Orac, VIENNA
Chile	Publitecsa and Conosur Ltda, SANTIAGO DE CHILE
Czech Republic	Orac sro, PRAGUE
France	Éditions du Juris-Classeur SA, PARIS
Hong Kong	Butterworths Asia (Hong Kong), HONG KONG
Hungary	Hvg Orac, BUDAPEST
India	Butterworths India, NEW DELHI
Ireland	Butterworths (Ireland) Ltd, DUBLIN
Italy	Giuffré, MILAN
Malaysia	Malayan Law Journal Sdn Bhd, KUALA LUMPUR
New Zealand	Butterworths of New Zealand, WELLINGTON
Poland	Wydawnictwa Prawnicze PWN, WARSAW
Singapore	Butterworths Asia, SINGAPORE
South Africa	Butterworth Publishers (Pty) Ltd, DURBAN
Switzerland	Stämpfli Verlag AG, BERNE
United Kingdom	Butterworths Tolley, a Division of Reed Elsevier (UK), LONDON, WC2A
USA	LexisNexis, DAYTON, Ohio

National Library of Canada Cataloguing in Publication Data

Keith, Norman, 1956-
    Workplace health and safety crimes : Bill C-45 and new Westray criminal offences / Norm Keith.

Includes bibliographical references and index.
ISBN 0-433-44473-8

    1. Criminal liability of juristic persons—Canada. 2. Industrial hygiene—Law and legislation—Canada. 3. Industrial safety—Law and legislation—Canada. 4. Corporation law—Canada—Criminal provisions. I. Title.

KE3365.K44 2004   345.71'04   C2004-901356-4
KE3570.K44 2004

Printed and bound in Canada

DEDICATION

For health and safety of young workers everywhere in Canada,
and especially Stephen, Rebekah, Rachel and Hannah.

ABOUT THE AUTHOR

Norm Keith is a partner at Gowling Lafleur Henderson, specializing in occupational health and safety law, workers' compensation, corporate liability, privacy law, labour and employment law and commercial litigation. He has extensive experience advising leading Canadian corporations on OHS risk management, in defending criminal and regulatory offences, wrongful dismissal litigation, breach of contract lawsuits, human rights complaints, labour arbitration hearings, judicial reviews, labour injunctions and fiduciary duty lawsuits. He has also assisted clients in a wide range of industries in developing their emergency response plans, drafting employment contracts, health and safety policies, human resource policies and violence and sexual harassment policies. Mr. Keith leads Gowlings' national OHS Team of lawyers and consultants providing multi-disciplinary training, consulting and legal services. Visit his website at *www.gowlings.com/ohscourses*.

A member of the Law Society of Upper Canada, Canadian Bar Association, Human Resources Professional Association of Ontario, Association of Canadian Registered Safety Professionals and Canadian Society of Safety Engineering, Mr. Keith has authored leading texts including *Canadian Health & Safety Law*, *Ontario Health & Safety Law*, *Creating an Effective Joint Health and Safety Committee* and *Human Resources Guide to Preventing Workplace Violence*. He was called to the bar in 1983 and received his Canadian Registered Safety Professional designation in 1998.

PREFACE

Bill C-45, *an Act to amend the Criminal Code*, is one of the most important legislative developments in recent Canadian legal history. Also known as the Westray Bill, Bill C-45 was the federal government's legislative response to the tragedy that occurred at the Westray mine in Pictou County, Nova Scotia. On May 9, 1992, 26 miners were killed when an explosion and fire ripped through the coal mine killing everyone who was underground in the mine at the time. There were legal, public inquiry and legislative responses to the Westray disaster; after much political and public pressure, the federal government responded. The Honourable Martin Cauchon, then Minister of Justice and Attorney General of Canada, who directed the introduction and passage of Bill C-45, made reference to the Westray mine disaster in a speech on June 12, 2003, when he introduced the bill for First Reading in the House of Commons.

Bill C-45 changes the landscape of occupational health and safety responsibility and potential liability in Canada. The *Criminal Code* applies to all Canadian workplaces. Unlike other Canadian health and safety laws, the Bill C-45 amendments create a legal duty to take "reasonable steps" in all workplaces across Canada. Further, the legislation creates a new crime of occupational health and safety criminal negligence. Indictable offences may result in individuals being sentenced to the maximum of life imprisonment and organizations, including corporations, may be fined in the tens of millions of dollars, or even higher.

Bill C-45, in addition to its occupational health and safety focus, also dramatically changes the law with respect to the criminal liability of organizations, including corporations, across Canada. Prior to this legislation, the *Criminal Code* was silent on the legal theory or means by which a corporation would be held criminally liable. The long-standing and much criticized "identification theory" of corporate criminal liability has now been replaced by two specific provisions in the *Criminal Code* to address the fault element of an organization, including a corporation. The first deals with offences that require proof of negligence and the second deals with those that require a more traditional, subjective fault element. These changes to the *Criminal Code* dramatically affect the criminal liability of organizations, including corporations, far beyond the new duty and crime with respect to occupational health and safety in the workplace.

This introductory text provides a background to the Westray disaster and its corresponding legal, public inquiry, and legislative responses. The book provides an introduction to criminal law in general and the law of criminal negligence in particular. To understand the new offence of occupational health and safety criminal negligence, one needs to understand the nature of the general offence of criminal negligence, as well as criminal procedure.

The book also offers a section-by-section analysis of Bill C-45. Each change to the *Criminal Code* is reviewed and discussed, providing insight and analysis on the language and potential application of each new provision. The book then provides an overview of the current state of occupational health and safety law in Canada, in particular, an introduction to the internal responsibility system, the external responsibility system and the legal defence of due diligence. To the surprise of many occupational health and safety professionals, human resource managers and even lawyers in the field, the legal defence of due diligence is not available to the new criminal offence of occupational health and safety criminal negligence. This point turns on the classification of the new offence. Several aspects of the Bill C-45 amendments are quite similar and analogous to the due diligence defence to occupational health and safety strict liability offences; however, the new workplace health and safety crimes established by Bill C-45 do not have a reverse onus on the accused.

Finally, in the post-Bill C-45 era of occupational health and safety criminal negligence, it is imperative for organizations, including corporations, and all those who direct how work is performed to be familiar with the criminal investigation and criminal prosecution process. This includes the right that both individuals and organizations, including corporations, have under our legal system and the *Charter of Rights and Freedoms*. The criminalization of breach of an occupational health and safety duty to take "reasonable steps" to protect workers and the public certainly heightens the importance of workplace accident and injury prevention. The management of individual and organizational risk requires the establishment of effective occupational health and safety programs and management systems.

I am proud to have established a national team of lawyers and occupational health and safety consultants at Gowling Lafleur Henderson that provides training, consulting, and legal services. This OHS Team assists clients across Canada to develop and implement the elements of an effective occupational health and safety management system that both reduces risk of accident injury on the job and ensures legal compliance.

I would like to acknowledge and thank the following articling and summer students at Gowling Lafleur Henderson who have assisted in providing research support for this book including Heather Gray, Julie Lavigne, Patricia Maclean, Goldie Bassi, Lindsey Cader, Francis Roy, Dominic Gregoire, Liza Swale, Jonathan Morse and Christine Goldsack. I would also like to thank my colleague, Yvonne O'Reilly, Gowlings' Senior OHS Consultant, for reading the entire manuscript and providing valuable suggestions to improve this book. I would also like to thank Eric Siebenmorgen, Ontario Crown Attorney, for his review and suggestions in respect of chapters 3 and 6, where his knowledge of criminal law procedure was most helpful.

Finally, I would like to thank my wife Diane for her continuing support and understanding in respect to the demands of my professional practice, and for managing her workplace with grace.

TABLE OF CONTENTS

ABOUT THE AUTHOR...v
PREFACE vii

1. INTRODUCTION...1
    Overview of this book...1
    Canadian criminal law..2
    Bill C-46 and corporate crime.......................................5
    The purposes of Bill C-45 7

2. BACKGROUND TO BILL C-45 15
    The Westray mine disaster 15
    The legal response...18
    The public inquiry...24
    Legislative responses 26
    Standing Committee on Justice and Human Rights 30
    Introduction of Bill C-45...33

3. THE LAW OF CRIMINAL NEGLIGENCE.......................39
    The Offence of criminal negligence.............................39
    Acts and omissions...42
    Objective vs. subjective standard.................................44
    Wanton or reckless disregard 47
    Corporate criminal liability 49
    The fault element for organizations and corporations.....51
    Models of corporate criminal liability..........................56

4. DETAILED ANALYSIS OF BILL C-45 63
    Section 1: Extends potential criminal liability 63
    Section 2: Extends potential criminal liability deeper into the
        organizational structure 65
    Section 3: Establishes a health and safety duty for persons
        directing work or authorized to do so. 68
    Section 4: Replaces "corporation" with "organization" with
        respect to the offence of theft. 73
    Section 5: Replaces "corporation" with "organization" with
        respect to false pretences. 74

Section 6: Repeals section 391 and amends subsection
418(2) of the *Criminal Code* 75
Section 7: Replaces "corporation" with "organization" with
respect to persons deemed to have absconded...........................76
Section 8: Replaces "corporation" with "organization" with
respect to procedure on preliminary inquiry...............................76
Section 9: Replaces "corporations" with "organization" with
respect to appearances and elections. 77
Section 10: Replaces "corporation" with "organization" with
respect to warrants of committal. 78
Section 11: Replaces "corporation" with "organization" with
respect to appearances and notice...79
Section 12: Replaces "corporations" with "organization" with
respect to presence in court during trial.......................................80
Section 13: Replaces "corporation" with "organization" with
respect to service of process...80
Section 14: Factors to consider when sentencing an
organization. 81
Section 15: Replaces "corporation" with "organization"
with respect to probation reports. 85
Section 16: Evidence of previous convictions 86
Section 17: Absolute and conditional discharges...........................86
Section 18: Extends definition of "optional conditions" of
probation...87
Section 19: Replaces "corporation" with "organization" with
respect to powers to impose fines...91
Section 20: Replaces "corporation" with "organization" and
increases limit of possible fines..92
Section 21: Replaces "corporation" with "organization" with
respect to appearances for summary conviction offences 93
Section 22: Coordinating amendment..94
Section 23: Coming into force date...94

5.   UNDERSTANDING CANADIAN OCCUPATIONAL
       HEALTH AND SAFETY LAW 97
Introduction to Canadian occupational health and safety
law 97
The Internal responsibility system 101
OHS law general duty clauses..104
Enforcement of OHS law and due diligence...................................108
OHS due diligence and Bill C-45 "reasonable steps" 118

6.   CRIMINAL CODE ENFORCEMENT PROCEEDINGS 121
     Introduction to the criminal judicial system.................121
     The classification of offences......................................123
     The right to be free from unreasonable search and
        seizure...125
     The right to be free from arbitrary detention..............127
     Initiation of criminal proceedings 127
     Arresting an accused 128
     Right to counsel 130
     Judicial interim release – a.k.a. "Bail".....................131
     Crown disclosure..133
     The accused's election in criminal negligence...........134
     The preliminary inquiry 136
     Pre-trial conference 137
     Trial..138
     Sentencing...141
     Some final thoughts 143

APPENDIX A Bill C-45 145
APPENDIX B Westray mine disaster — factual and legal
        chronology...155
APPENDIX C Consolidated recommendations of the Westray
        Inquiry 167

Index 185

CHAPTER 1

INTRODUCTION

OVERVIEW OF THIS BOOK

The purpose of this book is to help you understand Bill C-45. Bill C-45 has important implications for occupational health and safety ("OHS") law and the criminal liability of every individual who directs how others perform work, and every organization in Canada. Bill C-45 has often been referred to as the "Westray Bill' in the popular press and by those who have lobbied to both introduce and pass the bill. This reference is in memory of the 26 workers who died on May 9, 1992 at the Westray mine in Nova Scotia. Many personal memorials and memories exist with respect to the Westray disaster; the Westray disaster was a dark day in Canadian history. Bill C-45 has been promoted and passed into law with a view to protecting workers in Canadian workplaces and preventing another Westray disaster.

Bill C-45 is federal legislation that established an OHS duty on individuals, organizations and their decision-makers across Canada. Bill C-45, among other important changes to the *Criminal Code*, establishes OHS negligence as a criminal offence. The First Reading of Bill C-45 took place on June 12, 2003. It passed through the House of Commons and the Senate relatively quickly, receiving Royal Assent on November 7, 2003. It was proclaimed into force by an order of the Governor in Council and took effect on March 31, 2004. Maximum penalties for the new offence of OHS criminal negligence include life imprisonment for an individual and a fine with no maximum limit for an organization. This is the first time in Canadian history that a legal duty with respect to the health and safety of workers and the public has been included in the most important of Canadian penal statutes, the *Criminal Code*.

This book will provide an overview of and background to Bill C-45, an overview of criminal negligence, a detailed analysis of Bill C-45, an introduction to Canadian OHS law and an introduction to *Criminal Code* enforcement procedure. It also includes a discussion of the Westray mine disaster, the Public Inquiry of Justice Peter Richards which followed the disaster, the legal, political and legislative responses and recommendations from the Public Inquiry.

Bill C-45 fundamentally adds to, rather than replaces, existing provisions under the *Criminal Code*. For example, Bill C-45 replaces the old term "corporation" with the new term "organization". The term "organization" is broadly defined under the Bill C-45 amendments. The new liability provisions of Bill C-45 for organizations apply to more than just the new OHS criminal negligence offence. Bill C-45 dramatically amends the fault elements of all offences against organizations under the *Criminal Code*. Extensive new sentencing and probationary provisions introduced by the bill are also covered in this book.

A general overview of Canadian OHS law is important in order to understand the full implications of Bill C-45. This book includes a discussion of the internal and external responsibility systems in Canadian OHS law. Bill C-45 does not replace existing Canadian OHS statutes or regulations but it does add criminal liability to workplace health and safety. Bill C-45 does not provide additional requirements, hazard control measures, system criteria for an OHS management system but it does require all affected parties to take "reasonable steps" to prevent bodily harm. Bill C-45 increases the liability for OHS offences and provides criminal sanctions for breach of the new OHS legal duty.

The last chapter of the book provides an overview of the criminal justice system in Canada. That is the process by which criminal charges are prosecuted in Canadian courts. Bill C-45 increases the likelihood that more individuals and organizations will be exposed to the criminal justice system through charges of OHS criminal negligence. The police and local crown attorneys will investigate and prosecute offences under the *Criminal Code*. Bill C-45 is enforced by these *Criminal Code* procedures through the courts. The "new normal" after Bill C-45 has come into force is that individuals and organizations that neglect the safety of workers, and/or the public, may be subject to criminal prosecution. This chapter will help everyone understand their rights and the criminal justice process.

CANADIAN CRIMINAL LAW

Criminal law in Canada is legislated and applied in an increasingly complex social and legal framework. Canadian criminal law is enacted by the federal parliament primarily in the form of the *Criminal Code*.[1] Criminal law provides the most severe legal censure available in the Canadian legal system. The complexity of Canadian criminal law and the rights of the accused have been increased by the development of the Canadian *Charter of Rights and Freedoms*.[2] The *Charter of Rights and Freedoms* (the "*Charter*") is part of the constitution and the fundamental law of Canada.

Although the *Criminal Code* contains many specific criminal offences, they constitute only a small proportion of the total number of offences in Canada. Most penal offences in Canada are regulatory offences, also known as public welfare offences. These types may be legislatively enacted by the federal parliament, provincial parliaments or municipalities. Regulatory offences include highway traffic offences, environmental offences, offences for over-fishing, offences for harmful commercial practices including misleading advertising, and occupational health and safety ("OHS") offences.

The primary purpose of regulatory statutes and offences is to promote the public good, deter risky behaviour, prohibit anti-social behaviour, and prevent all manner of social harm. The contravention of regulatory statutes, the resulting prosecution and conviction of a regulatory offence results in penalties such as a fine, monetary penalty and, in some instances, imprisonment.

Criminal offences in Canada generally have two elements. The first element is often referred to as the prohibited or unlawful act, or *actus reus*. It is well established in criminal law that the crown prosecutor, prosecuting on behalf of the people and government of Canada, must prove, beyond a reasonable doubt, that the accused has committed the prohibited act. The specific language that defines and establishes the prohibited act is found in the *Criminal Code* or other criminal statute.

The second element of a criminal offence is the mental intent or fault element, also known as *mens rea*. The crown prosecutor must also prove, beyond a reasonable doubt, this second element of a criminal charge to result in a criminal conviction. Mental intent or the fault element of the crime may be either inferred in the language of the criminal offence or specifically described in the language of the criminal offence. The fault

---

[1] R.S.C. 1985, c. C-46.
[2] *Canadian Charter of Rights and Freedoms* Part I of the *Constitution Act*, 1982 being Schedule B to the *Canada Act* 1982 (U.K.), c. 11 ("The *Charter*") came into effect on April 1, 1982. It provides accused individuals, and in some cases organizations and corporations, with a variety of legal rights including the right to fundamental justice and the right to a fair trial.

element may be specifically included in the language of the offence by use of language such as "intentional", "knowingly", "recklessly", or "negligently". The fault element of a crime is not present in most regulatory offences. The three categories of offences in Canadian law, including criminal offences and strict liability offences, are reviewed later in the book.

If the crown prosecutor has proven both the prohibited act and the fault element beyond a reasonable doubt, the accused is normally convicted. A number of legal defences are available to an accused in Canadian criminal law. Defences may vary in their nature and legal basis. For example, some defences may excuse or justify conduct even though there is proof beyond a reasonable doubt that the accused committed the offence with the necessary fault element. Some defences, such as mistake of fact and intoxication, are not technically defences at all, but rather an answer to the charge that would prevent the crown prosecutor from proving the required fault element. The failure of the crown prosecutor to prove both elements of the criminal offence will have the same effect as a positive defence, namely a finding of not guilty and an acquittal of the charges.

Criminal charges may lead to either a plea of guilty, or a plea of not guilty, resulting in a trial. The accused has the right to plead either guilty or not guilty. In a trial, the crown prosecutor proceeds first, calls witnesses, presents documents, and by other lawful means tenders evidence to prove its case beyond a reasonable doubt. At the end of the prosecutor's case, the defence lawyer representing the accused has an opportunity to call evidence through witnesses, documents, and other lawful means to advance the defence of the accused. At the end of the trial, a judge or a jury, as the case may be, will make a final decision with respect to guilt. The accused, in the Canadian criminal justice system, may be found either guilty or not guilty. If there is a finding of not guilty, an acquittal will be entered and, subject to the crown prosecutor's right of appeal, that is the end of the criminal charge. If there is a finding of guilt, the trial process will then result in a sentencing hearing to determine the appropriate penalty.

Sentencing of an accused takes place after the finding of guilt. The majority of criminal charges laid in Canada each year result in guilty pleas. After a guilty plea, or a conviction after a trial, there will be a sentencing hearing. The *Criminal Code* prescribes the maximum penalty available to the court imposed against the accused for each criminal offence. Judges have significant sentencing discretion in Canadian criminal law. Most criminal offences do not have a minimum punishment. There is, however, a trend toward setting legislatively mandated minimum penalties for various criminal offences. A primary consideration in the sentencing of an accused who has been convicted of a criminal offence is that the punishment must be proportionate to the seriousness of the offence. The more serious the criminal offence for which an accused has been convicted, generally the more onerous the criminal law penalty that will be imposed.

BILL C-46 AND CORPORATE CRIME

On June 12, 2003, the same date that Bill C-45 was given first reading in the House of Commons, the Minister of Justice and Attorney General for Canada also introduced Bill C-46,[3] an Act to amend the *Criminal Code*, (capital markets fraud and evidence-gathering), as a companion bill to Bill C-45. Martin Cauchon, then Minister of Justice and Attorney General of Canada, introduced Bill C-46 to establish a number of new crimes as part of the response to the Enron and WorldCom scandals of the U.S., insider trading charges such as those against Michael Cowpland, the co-founder of Corel, a Canadian software company, and Martha Stewart, the modern epitome of gracious living. Bill C-46 was intended to increase organizational criminal liability for financial sector fraud.

Bill C-46 had nothing to do with occupational or public safety. This bill, however, like Bill C-45, proposed new crimes that had not previously been in the *Criminal Code*. One provision of Bill C-46 stated:

> Every one who, by deceit, falsehood or other fraudulent means, whether or not it is
> a false pretence within the meaning of this Act, with intent to defraud, affects the
> public market price of stocks, shares, merchandise or anything that is offered for
> sale to the public is guilty of an indictable offence and liable to imprisonment for a
> term not exceeding fourteen years.[4]

Provincial securities legislation has already established rules against insider trading and regulatory offences for those who participated in insider trading. Insider trading is the activity by which a person with special information that may affect the valuation of securities trades on that information or knowledge before it is made available to the public. Bill C-46 proposed to add a new section to the *Criminal Code*, to

---

[3]   Bill C-46, *An Act to amend the Criminal Code*, 2nd Sess., 37th Parl., 2003, (as passed
      by the House of Commons 5 November 2003), online: <http://www.parl.gc.ca/37/2/
      parlbus/chambus/house/bills/government/C-46/C-46_3/C-46_cover-E.html>

[4]   *Ibid.*, s. 2(2).

establish a new crime of insider trading. Subsection 382.1(1) stated the following:

> A person is guilty of an indictable offence and liable to imprisonment for a term not exceeding ten years who, directly or indirectly, buys or sells a security, knowingly using inside information that they
>
> (a) possess by virtue of being a stockholder of the issuer of that security;
>
> (b) possess by virtue of, or obtained in the course of, their business or professional relationship with that issuer;
>
> (c) possess by virtue of, or obtained in the course of, a proposed takeover or reorganization of, or amalgamation, merger or similar business combination with, that issuer;
>
> (d) possess by virtue of, or obtained in the course of, their employment, office, duties or occupation with that issuer or with a person referred to in paragraphs (a) to (c); or
>
> (e) obtained from a person who possesses or obtained the information in a manner referred to in paragraphs (a) to (d).[5]

Bill C-46 also proposed to establish an anti-reprisal provision preventing employers from taking retaliatory action against any employee because the employee has provided information to a federal or provincial law enforcement agency with respect to a breach of the provisions under Bill C-46,[6] broad powers to courts to produce documents or data as ordered by a court,[7] give power to a court to order a financial institution to produce the account number of a person under an order by the court in relation to these new amendments to the *Criminal Code*.[8]

Bill C-46 did not receive the same broad political support and public attention that Bill C-45 received. For example, the United Steel Workers of America union and the victims of the Westray mine disaster neither lobbied for nor publicly commented on Bill C-46 as they did for Bill C-45. However, the Certified General Accountants Association of Canada issued a public statement of concern about Bill C-46. In an open letter to the Minister of Justice and Attorney General of Canada, the president and chief operating officer of the Certified General Accountants Association of Canada said, "The primary concern of CGA-Canada relates to the requirement that identified persons produce confidential or

---

[5] *Ibid.*, s. 5.
[6] *Ibid.*, s. 6.
[7] *Ibid.*, s. 7.
[8] *Ibid.*

privileged information or documents pursuant to an *ex parte* production order. In particular, CGA-Canada is concerned that accountants in possession of confidential or privileged information or documents may have to produce this material to an investigating authority before they can claim an exemption from the requirement to disclose."[9]

Although Bill C-46 did not obtain swift legislative passage prior to Prime Minister Chretien's resignation, as Bill C-45 did, it was clearly part of a broader initiative by the federal government to deal with issues relating to corporate governance and corporate crime. The successful passage of Bill C-45, in contrast to the unsuccessful movement of Bill C-46 through the parliamentary review and approval process, may indicate several things. First, Bill C-45 had strong public and political support and Bill C-46 did not. Second, the period of turmoil prior to the resignation and retirement of Prime Minister Chretien in the fall of 2003 clearly resulted in very little legislative business being conducted in the House of Commons and Senate of Canada. Third, occupational and public safety in Canada was given legislative priority over an initiative to increase the accountability of senior executives and directors, and others, in respect of financial and corporate governance reform.

THE PURPOSES OF BILL C-45

In order to assess the impact and effect of Bill C-45, one of the appropriate measures is to first determine the purposes of the bill and then to determine if they have been achieved. There are several means by which the purposes of new legislation can be determined: first, from the statements of intention by the government; second, the statements by, in this case, the former Minister of Justice and Attorney General for Canada, Martin Cauchon and others in the legislature, related speeches and other analysis; third, by a careful and detailed analysis of the legislation itself. The first two of these approaches to determine the purpose of the legislation are briefly reviewed below. The third is the subject matter of chapter 4.

Prior to reviewing some of the public statements and legislative debates with respect to Bill C-45, it is important to understand their legal significance. As courts may consider and interpret the Bill C-45

---

[9] Letter from Anthony Ariganello, CGA, President and Chief Operating Officer, Certified General Accountants Association of Canada, in an open letter dated October 31, 2003 to The Honourable Martin Cauchon, Minister of Justice and Attorney General of Canada, online: <http://www.cga-canada.org/eng/news/letter_bill_c46_e.htm>.

amendments to the *Criminal Code*, the question may arise as to whether or not the courts may use political statements and legislative debates as an aid to interpret this new legislation. Canadian courts have been traditionally reluctant in admitting and relying upon political statements and parliamentary debate in interpreting statutes in the course of their legal enforcement. For example, in the case of *Canada (Attorney General) v. Reader's Digest Assoc. Can. Ltd.*,[10] the Supreme Court of Canada refused to admit statements made by two Ministers in the House of Commons. In a later case, Justice Dickson of the Supreme Court of Canada stated, "Generally speaking, speeches made in the legislature at the time of enactment of the measure are inadmissible as having little evidential weight".[11] More than ten years later, the Supreme Court of Canada, in a constitutional challenge, somewhat relaxed its approach for the purpose of constitutional litigation. In *R. v. Morgentaler*, Justice Sopinka said, speaking for the court, "Provided that the court remains mindful of the limited reliability and weight of Hansard evidence, it should be admitted as relevant to both the background and the purpose of legislation."[12]

In constitutional challenges in the courts the legislative history, including political statements and legislative debates, may be considered by a court in interpreting the legislation whose constitutionality is being challenged. In ordinary statutory interpretation, political statements and legislative debates may only be used to determine the specific purpose or mischief that parliament sought to address and correct by passing the legislation in question. Generally speaking, political statements and legislative debates may not be used to determine or interpret the specific meaning of a particular provision. Therefore, Bill C-45 may be better understood by the political statements and legislative debates, but it will not likely have the interpretation of its specific statutory provisions affected by political statements and legislative debates.

The purposes of Bill C-45 can be determined by information provided by the government at the time of and after the presentation of legislative amendments for parliamentary review, debate, and passage. In that respect, the government of Canada prepared a Plain Language Guide to Bill C-45 that is reviewed below. Further, the parliamentary research

---

[10] [1961] S.C.R. 775.

[11] *Reference re: Residential Tenancies Act 1979 (Ontario)*, [1981] 1 S.C.R. 714, Dickson J.

[12] [1993] 3 S.C.R. 463, Sopinka J.

branch prepared a legislative summary and analysis of Bill C-45. These sources are instructive with respect to the purposes of Bill C-45.

Four purposes of Bill C-45 are set out in the summary that introduces the legislation. The full text of the bill is contained in Appendix "A" to this book. The bill amended the *Criminal Code* to (a) establish rules for attributing to organizations, including corporations, criminal liability for the acts of their representatives, (b) establish a legal duty for all persons directing work to take reasonable steps to ensure the safety of workers and the public, (c) set out factors for courts to consider when sentencing an organization, and (d) provide optional conditions of probation that a court may impose on an organization.

Further, in a press release issued on June 12, 2003, the day Bill C-45 was introduced in Parliament, the Honourable Martin Cauchon, then Minister of Justice and Attorney General of Canada, said:

"Employers must fully recognize their responsibility in providing a safe work environment. Failure to do so in a manner that endangers employee and public safety must be appropriately dealt with through our criminal laws. I am pleased to introduce measures today that will effectively modernize the law on corporate liability."[13]

At Bill C-45 second reading debate on September 15, 2003 in the House of Commons, Mr. Paul Harold Macklin, Parliamentary Secretary to the Minister of Justice and Attorney General of Canada, said:

> Fundamentally the bill has its origins in the tragic deaths of 26 miners in the Westray mine explosion in May 1992. I will not review in detail the lengthy and ultimately fruitless criminal proceedings that followed the investigation of the explosion. All members are aware that the company that operated the mine, and two of its executives, were charged with manslaughter. The trial judge ordered a stay of the charges because of problems with disclosure of evidence by the Crown. Although the appeal courts overturned that decision, the prosecution decided it could not go forward.[14]

Mr. Macklin described the results of the Public Inquiry into the Westray disaster and the growing consensus in respect of the need to reform the criminal law relating to the legal liability of corporations. He then described the role of the Standing Committee on Justice and Human

---

[13]   Department of Justice Canada, News Release, "Justice Minister Introduces Measures to Protect Workplace Safety and Modernize Corporate Liability" (12 June 2003), online: <http://canada.justice.gc.ca/en/news/nr/2003/doc_30922.html>.

[14]   *House of Commons Debates*, 119 (15 September 2003) at 1340 (P.H. Macklin), online: <http://www.parl.gc.ca/37/2/parlbus/chambus/house/debates/119_2003-09-15/HAN119-E.HTM>.

Rights in the consensus building process that resulted in Bill C-45. He said:

> A discussion paper setting out the issues and reviewing the evidence of other countries, which had been prepared by the justice department, was provided to the committee. The committee heard from officials of the justice department and other experts. It heard moving testimony from victims and relatives of victims of industrial accidents. The 15th report of the committee recommended "that the government table in the House legislation to deal with the criminal liability of corporations, directors and officers". Clearly all parties in the House felt that it was time for fundamental reform in this area. The government in its response to the report reviewed the evidence that had been heard by the committee and agreed on the need for reform. The government also concluded that there was no perfect system in other countries that Canada could simply copy. The report therefore set out the principles that would guide the drafting of a made in Canada approach to the problem of corporate crime.[15]

He went on to describe what was characterized as an innovative step of introducing a new OHS duty in the *Criminal Code*. Although Mr. Macklin reminded the House of Commons that the federal government had recently amended the *Canada Labour Code,* Part II, to increase employer responsibilities, employee rights, and the fines associated with the contravention of federal *OHS* legislation, he indicated that new *Criminal Code* amendments were also required. He went on to say:

> Bill C-45 builds on those changes by proposing to include in the *Criminal Code* a new section, section 217.1, which provides that everyone who undertakes, or has the authority, to direct how another person does work or performs a task is under a legal duty to take reasonable steps to prevent bodily harm to that person or any other person arising from that work or task. The importance of having such a duty in the *Criminal Code* is that if there is a breach of that duty, wanton and reckless disregard for the life or safety of people, and injury or death results from that breach, a person can be convicted of criminal negligence causing death which is punishable by up to life imprisonment, or criminal negligence causing bodily harm which is punishable by up to 10 years imprisonment.[16]

The Department of Justice of the government of Canada also prepared and provided a Plain Language Guide to Bill C-45.[17] The government's Plain Language Guide gave further background on the purposes of Bill C-45. It said:

---

[15] *Ibid.* at 1335.

[16] *Ibid.* at 1340.

[17] Canada, Department of Justice, *A Plain Language Guide, Bill C-45 – Amendments to the Criminal Code Affecting the Criminal Liability of Organizations*, online: <http://canada.justice.gc.ca/en/dept/pub/c45/index.html>.

> The government tabled Bill C-45…to modernize the law with respect to the criminal liability of corporations and the sentencing of corporations…[b]ecause the *Criminal Code* covers a wide range of crimes by all kinds of persons, the legislation employs more complex and specific language…[18]

The Plain Language Guide went on to say, "the provisions of Bill C-45 are a compilation of the existing rules with new reforms, which will modernize the law to reflect the increasing complexity of corporate structures."[19]

The emphasis in the Plain Language Guide is on the change that Bill C-45 makes to the criminal liability of organizations. What is striking by its absence is that there is no specific reference to the new legal OHS duty in the Plain Language Guide. However, it does provide an interesting example of how an organization may be held liable for OHS criminal negligence in the post-Bill C-45 era of the *Criminal Code*. The Plain Language Guide offers the following example of how an organization will be held accountable:

> For example, in a factory, an employee who turned off three separate safety systems would probably be prosecuted for causing death by criminal negligence if employees were killed as a result of an accident that the safety systems would have prevented. The employee acted negligently. On the other hand, if three employees each turned off one of the safety systems each thinking that it was not a problem because the other two systems would still be in place, they would probably not be subject to criminal prosecution because each one alone might not have shown reckless disregard for the lives of other employees. However, the fact that the individual employees might escape prosecution should not mean that their employer necessarily would not be prosecuted. After all, the organization, through its three employees, turned off the three systems…Similarly, in the example of three employees engaging in the negligent conduct, the court would have to decide whether the organization should have had a system to prevent them from acting independently in a dangerous way and whether the lack of such a system was a marked departure from the standard of care expected in the circumstances. The court would consider, under this example, the practices put in place by the person in charge of safety at the factory and the practices of other similar organizations.[20]

The Library of Parliament, through its parliamentary research branch, prepared a legislative summary of Bill C-45.[21] Although this legislative summary was prepared in the summer of 2003, after the first reading of

---

[18]   *Ibid.* at 1.

[19]   *Ibid.*

[20]   *Ibid.*

[21]   Parliamentary Research Branch of the Library of Parliament, *Bill C-45: An Act to Amend the Criminal Code (Criminal Liability of Organizations)* (Legislative Summary) David Goetz (Ottawa: Law and Government Division, 2003).

Bill C-45, it is still helpful and relevant since there were very few changes to Bill C-45 in the legislative review and debate process. In providing background to Bill C-45 the subject of the identification theory of corporate criminal liability was reviewed.[22] The legislative summary stated, in that regard:

> The "identification theory" of corporate criminal liability has been criticized as inadequate over the years, both in Canada and elsewhere. Critics of this approach have pointed out that it does not reflect the reality of the internal dynamics of corporations, particularly in the case of larger corporations. Rarely do high-level corporate officials personally engage in the specific conduct or make the specific decisions that result in occupational health and safety violations or in serious workplace injury or death. However, they can often, through actual policy decisions or otherwise, create or contribute to a corporate environment where subordinate managers, supervisors and employees feel encouraged or even compelled to cut corners on employee health and safety matters, even in the face of legal prohibitions or official corporate policy.[23]

The Parliamentary Research Branch legislative summary of Bill C-45 made a passing reference to the Westray mine disaster. Surprisingly, even though the lobbying efforts from the United Steel Workers of America union, families and relatives of the Westray mine victims spearheaded the push for what became Bill C-45, the Westray disaster is not significantly recognized in the legislative summary. Further, there is no mention of the Westray disaster in the Plain Language Guide, prepared by the federal government. The legislative summary's brief reference to the Westray disaster is as follows:

> In its submissions to the public inquiry into the Westray mine disaster of May 1992, the United Steel Workers of America called for the facilitation of corporate criminal liability and also advocated enhanced criminal accountability of corporate directors and officers...the union recommended creating a new criminal offence aimed specifically at "directors and responsible corporate agents" who negligently fail to protect the health and safety of employees. The union conceded that the offence would likely have to be confined to situations of *criminal* negligence (i.e., conduct amounting to a "marked departure" from the standard of reasonable person). However, it was thought that legislating a specific legal duty on the part of key corporate officials to take reasonable care to protect employees would facilitate their prosecution by obviating the need to establish a causal connection between the conduct of a corporate official and the death or injury of an employee.[24]

---

[22] For more information on the 'identification theory' of corporate criminal liability see Chapter 3.

[23] *Bill C-45*, Legislative Summary at 5.

[24] *Ibid.* at 6.

The legislative summary goes on to provide a brief overview of various provisions of Bill C-45, without further elaborating on the purpose of the bill. Interestingly, in the 14 page legislative summary, only one relatively short paragraph deals with the new OHS legal duty established by Bill C-45. That excerpt states, in the writer's view rather inaccurately, the following description of new section 217.1 of the *Criminal Code*:

> Clause 3 of the bill amends the *Criminal Code* by adding a new section 217.1 which will provide that those who are responsible for directing the work of others are under a legal duty to take reasonable steps to prevent bodily harm to any person arising from such work. This provision does not create a new criminal offence. However, by clarifying the existence of such a legal duty, the provision facilitates the application of the offence of criminal negligence, which is predicated, in part, on the existence of a legal duty.

This description of the new OHS legal duty in the *Criminal Code* appears to diminish the importance of the workplace health and safety aspect of Bill C-45. In the writer's view, Bill C-45 not only establishes a new OHS legal duty but it also establishes a new health and safety crime, OHS criminal negligence. Further, it would appear that the legislative summary and the Plain Language Guide to Bill C-45 are preoccupied with the change to criminal liability of organizations to the detriment of the OHS portion of the bill. Both government documents are deficient in their review, analysis and implications of the new OHS legal duty in section 217.1 of the *Criminal Code*. The fact that Bill C-45 creates a new offence of OHS criminal negligence, for breach of the new OHS legal duty in the *Criminal Code*, is clearly one of the central purposes and results of Bill C-45. Both the bill itself and the Minister who introduced it to Parliament said that Bill C-45 is intended to address workplace health and safety for the benefit of workers and the public by establishing this new criminal accountability.

CHAPTER 2

BACKGROUND TO BILL C-45

THE WESTRAY MINE DISASTER

To understand the historical and political impetus for Bill C-45 one must be aware of the Westray mine disaster. On May 9, 1992, at approximately 5:20 a.m., 26 miners in Plymouth, just east of Stellerton, in Pictou County, Nova Scotia, were killed when an explosion, fire and ground collapse occurred at the Westray coal mine. There were no survivors of the Westray disaster.

Tragically, this was neither the only nor the worst coal mine disaster in the coal mining area of Nova Scotia known as the Foord Coal Seam. In 1873, an explosion and fire killed 60 workers at the Drummond Mine in Westville Nova Scotia; in 1880 an explosion resulted in the death of 44 miners at the Foord Pit; in 1818, there was an explosion and fire at the Alan shaft that resulted in the death of 88 miners.[1] Coal mining is a hazardous business.

In the wake of the Westray tragedy, many questions remain. First and foremost was the ultimate question of exactly why the explosion occurred. Much information, evidence and speculation came to light in the aftermath of the Westray disaster. Ultimately, reliable evidence demonstrated that there was a "complex mosaic of actions, omissions, mistakes, incompetence, apathy, cynicism, stupidity, and neglect"[2] that contributed to the disaster. Westray mine managers' officials clearly did not follow proper safe mining procedures.

The Westray mine was developed by the corporation Curragh Inc. ("Curragh") with financial support of both the Nova Scotia provincial and federal government. Among other agreements, Westray had negotiated a 15-year contract with the Nova Scotia Power Corporation

---

[1] Shaun Comish, *Westray Tragedy: A Miner's Story* (Halifax: Fernwood Publishing, 1993) at 2.

[2] Justice K. P. Richard, Commissioner, *The Westray Story: A Predictable Path to Disaster*, Executive Summary (Pictou: Westray Mine Public Inquiry, 1997) at vii.

("NSPC") to supply more than 275,000 tonnes of coal per year.[3] The mine, when it opened in 1991, was lauded as a boon to the local economy of Pictou County, promising steady employment in a region where unemployment was high.

Curragh had never developed an underground coal mine prior to Westray. It did, however, have experience in mining base metals in western Canada, and it was a large producer of lead and zinc.[4] Curragh would be paid $60 per tonne from the NSPC and an additional $14 per tonne for premium-grade coal, which is higher in heat content and lower in impurities. Selling lower-grade coal would provide Westray with roughly $42 million in annual revenues,[5] and almost $50 million in annual revenues if the target of shipping more than 50% of its output as premium coal was met. With projected operating costs of $35 million, a profit of between $7 million and $15 million was estimated.[6]

There is little doubt in retrospect that these production figures were too optimistic. In developing its plans, Curragh relied heavily on various feasibility and planning studies that had been prepared for other companies which were initially interested in obtaining the mining lease to the area. It is possible that if Curragh had approached the development of the Westray mine with higher standards of worker safety, to eliminate or better control potential workplace risks and dangers, the business case for going forward with the project would not have been as strong.

Some of the hazards of the Westray mine were later described by one of the miners who worked in the mine up to the time of the explosion:

> I can't count how many times cave-ins buried the front of the miner [large piece of mining equipment]. When all the rocks came crashing down, it would shower me with small pieces of rock that flew through the protective screening. I would have to duck my head until all the debris stopped flying. Once the dust settled, I would look around and assess the situation. Most times I could back the miner out and clean up the mess, but once in a while I got stuck…you were always wondering when the roof was going to come crashing down.…[7]

About 225 people in total worked at the Westray mine during peak production times. Occasional concerns were raised about the safety of the mine, but the work was not stopped. Many of the miners at Westray

---

[3]   *Ibid.* at 4.
[4]   Dean Jobb, *Calculated Risk – Greed, Politics, and The Westray Tragedy* (Halifax: Nimbus Publishing Ltd., 1994) at 16.
[5]   *Ibid.*
[6]   *Ibid.*
[7]   Shaun Comish, *Westray Tragedy: A Miner's Story.*

were second or third generation miners. In a coal mining region such as Pictou County, danger and death were all too familiar to the community:

> From 1866 to the early 1970s, 576 men and boys died in Pictou's collieries. Of that number, 246 perished in methane and coal-dust explosions. The remaining 330 fell victim, one or two at a time, to other underground hazards. Many were crushed in cave-ins; others were suffocated by gases, run over by coal cars, or mangled after becoming caught in machinery. Those numbers do not tell the whole story; for the first fifty-seven years after commercial production began in 1809, no records were kept.[8]

Furthermore, in an economically depressed region such as Pictou County of Nova Scotia, it had been historically easier to overlook obvious dangers at work when the alternative was unemployment. The dangerous nature of coal mining, however, mixed in with a poor attention to workplace health and safety proved to be a disastrous combination.

The need to have a good paying job was strong incentive for miners to continue to work even though they had health and safety concerns. The challenge of finding sustained full-time work in this region of Nova Scotia was immense. One miner who worked at Westray expressed his feelings on this subject as follows:

> A lot of people ask me why we kept working there. I guess the only answer I can give is that nowadays when you have a job it is very scary to quit and hope you can get a job somewhere else...some guys who worked at Westray didn't really know anything else but mining...the promise of fifteen years of steady work weighed heavy on your mind.[9]

The May 9, 1992 explosion at the Westray mine occurred as a result of a build-up of methane gas in one of the mine sections, which then was ignited by a spark, possibly from one of the cutting machines.[10] Methane gas is a naturally occurring, but highly flammable, by-product in the mining of coal. Its excessive accumulation can usually be controlled with proper ventilation systems. Extremely high concentrations of methane gas would make it impossible for a worker to breathe since the high levels are an indication of correspondingly low oxygen levels in the mine. Methane gas in low concentrations, on the other hand, creates a condition that is ripe for an explosion. This is especially true where, as in

---

[8]   *Ibid.*

[9]   *Ibid.*

[10] While the exact cause of the Westray explosion was never definitively determined, this is the most likely conclusion, as reached by Justice Richard in his report, *The Westray Story: A Predictable Path to Disaster.*

the Westray mine, coal dust is allowed to build up throughout the mine. Proper ventilation systems in a coal mine help to keep methane gas at a manageable level. When methane gas ignites, if coal dust is also present in high levels, a fire or explosion that might otherwise be isolated to one area can literally roll its way through an entire mine.[11] This is what many witnesses at the Westray Public Inquiry testified likely caused the explosion and resulting fire at Westray.

Numerous Nova Scotia government mine safety inspector reports warned of problems involving potentially explosive coal dust, cave-ins and dangerous levels of methane gas at Westray. Government safety inspectors requested Westray management to correct the problem, either through proper ventilation to reduce the presence of methane gas or through the spreading of stone dust to reduce the flammability of coal dust.[12] Westray management took some steps but also offered excuses for delay in implementing these proper workplace health and safety measures.

The government safety inspectors, under provincial legislation, did have the legal authority to ensure that safety procedures were followed at Westray. The power to close down a mine for safety reasons if necessary also existed under the *Coal Mines Regulation Act*.[13] Neither step was ever taken by the government of Nova Scotia OHS regulator. One of the many tragedies of the Westray disaster is that the Nova Scotia government regulators did not rigorously enforce existing OHS regulatory requirements.

THE LEGAL RESPONSE

Just as a complex of factors contributed to the Westray mine disaster, the legal response to the death of the 26 miners in Pictou County was complex. The initial legal responses to the Westray mine disaster resulted in investigations and prosecutions by both the Nova Scotia Department of Labour and Royal Canadian Mounted Police. The chronology of the OHS charges as well as criminal charges is summarized in Appendix B to this book.

---

[11] *Ibid.* at 209.
[12] Dean Jobb, *Calculated risk—Greed, Politics, and The Westray Tragedy* (Halifax: Nimbus Publishing Ltd., 1994) at 6-7. See also Canada, House of Commons, Standing Committee on Justice and Human Rights, "*Corporate Liability*", online: <www.parl.gc.ca/InfoComDoc/37/1/JUST/Studies/Reports/JustRP15-E.htm>.
[13] R.S.N.S. 1989, c. 73, ss. 63(1)(e), 64(1), as rep. by *Occupational Health and Safety Act,* 1996, S.N.S., c. 7, as am. by S.N.S. 2000, c. 28, ss. 86-87.

In addition to those prosecutions, the Public Inquiry that was first promised on May 10, 1992, one day after the Westray disaster, also resulted in various, complex legal challenges and controversy. In fact, shortly after the Public Inquiry (the "Inquiry") was called, and Premier Donald Cameron appointed Mr. Justice Peter Richard, of the Nova Scotia Supreme Court, to carry out the Inquiry, it was the subject of a judicial application for an interim injunction. On September 30, 1992, Gerald J. Phillips, Roger Parry, Glyn Jones, Arnold Smith, Robert Parry, Brian Palmer and Kevin Atherton were successful in obtaining a temporary injunction against the commencement of the Public Inquiry to the Westray disaster.[14] The applicants challenged the constitutionality of the Westray Inquiry on a number of grounds, including a concern about potential criminal charges, and the scope and authority of the Inquiry. His Honour Judge Glube, Chief Justice of the Trial Division of the Nova Scotia Supreme Court, characterized the application as follows:

> The applicants, in their main application, challenge the constitutionality of the Inquiry on the basis that it is *ultra vires* the province, because it is usurping the function of a section 96 court by encroaching on the federal criminal law and procedures, and, also, that section 67 of the *Coal Mines Regulation Act* is *ultra vires*, again, for the same reasons, and further, that the Inquiry violates the constitutional rights of the applicants under certain sections of the *Charter*.[15]

When Justice Glube granted the interim injunction and stopped the Inquiry from proceeding, no charges under either the Nova Scotia *Occupational Health And Safety Act* or the *Criminal Code* had been laid. The court went on to state:

> The Inquiry must be legally constituted before proceeding, in my opinion. In that way, the public interest will be served because, although there will be a delay in knowing the whys and answers to all of the questions, it will be a valid Inquiry that proceeds, after a court determination. In my opinion, the public interest as well as the private interest is best served by granting a temporary stay.[16]

The legal challenges with respect to the appropriateness of the Inquiry commencing with the possibility of OHS regulatory and criminal charges pending, wound its way slowly to the Supreme Court of Canada. On May 4, 1995, the Supreme Court of Canada granted an appeal from the decision of the Nova Scotia Court of Appeal, which had ordered that the Inquiry be stayed pending the resolution of the charges, which had

---

[14] *Phillips v. Nova Scotia (Commission of Inquiry into the Westray Mine Tragedy)* (1992), 116 N.S.R. (2d) 30 (N.S.S.C.).
[15] *Ibid.* at 32.
[16] *Ibid.* at 33.

subsequently been laid, under the *Criminal Code* against the individual accused.[17] The Supreme Court of Canada noted that the Court of Appeal for Nova Scotia had ordered the stay of the public hearing on the basis that although the individual accused were not compellable witnesses at the Inquiry, nevertheless, the publicity that would likely be generated by the conduct of the Inquiry would result in the infringement of the accused's rights under section 7 of the *Charter* by reason of the effect of the publicity on jurors. The Supreme Court of Canada held that since the accused had since re-elected trial by judge alone, without a jury, it was unnecessary to further consider the issue. This was the majority view of five of the nine justices of the Supreme Court of Canada. The decision to permit the Inquiry to proceed therefore, was based on very narrow grounds.

However, Justice Cory of the Supreme Court of Canada, with Justices Iacobucci and Major, also agreed that the Inquiry should proceed but gave more consideration to the public policy considerations. Justice Cory stated that a balance must be struck between the state's interest in obtaining evidence for a valid public purpose and an individual's right to remain silent and to have a fair trial. The crucial issue for the court to determine, according to Justice Cory, was the question of what is the predominant purpose for seeking the evidence at the Inquiry. If the evidence at the Inquiry was primarily to obtain evidence for prosecution of a witness, section 77 of the *Charter* would require that the accused person be exempted from testifying. If the primary purpose for an Inquiry was to look into the cause of the disaster, then the witness may not have such an exemption. Further, Justice Cory went through a lengthy analysis on how to protect the rights of an individual, assuming there was still a jury trial of the criminal charges, even if the primary purpose was to identify causes of the explosion. This analysis by Justice Cory may assist courts and future public inquiries and coroners' inquests in dealing with this issue. However, since it was not the majority decision of the Supreme Court of Canada, it has generally been regarded as non-binding.

Charges under Nova Scotia's then *Occupational Health and Safety Act* were laid on October 5, 1992. However, apparently out of caution leading to some overlap with the criminal charges that were later laid, 32 of the 52 charges were voluntarily stayed or dropped by the Department of Labour in December 1992. In March 1993 the remaining charges

---

[17] *Phillips v. Nova Scotia (Commission of Inquiry into the Westray Mine Tragedy)* (1995), 124 D.L.R. (4th) 129 (S.C.C.).

under the *Occupational Health and Safety Act* were dropped by the Nova Scotia Department of Labour. There never was a trial of the OHS regulatory charges under Nova Scotia's *Occupational Health and Safety Act* relating to the Westray mine disaster.

On May 21, 1992, the Royal Canadian Mounted Police launched a criminal investigation of the Westray mine disaster, initiated partly as a result of allegations that documents were being shredded at the Westray mine site. On April 20, 1993, Curragh, Gerald Phillips, the Westray mine manager, and Roger Parry, the Westray underground manager, were charged with manslaughter and criminal negligence causing death under the *Criminal Code*. The validity of these criminal charges was challenged by the accused. On July 20, 1993, provincial court Judge Curran granted a motion to stay all of the charges, subject to the right of the Crown to lay new charges against the three accused.[18] The three accused moved to have the charges against them stayed on the grounds that they lacked sufficient particularity and failed to identify the transactions giving rise to the allegations and the charges. Justice Curran stated, making his decision:

> Accordingly, because the counts in the information before the court totally fail to describe the transactions alleged to have constituted the offences, and because I have no power at this stage to amend or to order particulars, both counts are quashed and the accused are discharged. The Crown, of course, has the right to lay new charges.[19]

New criminal charges were subsequently laid against the same three accused. Those charges were also subject to a similar motion to quash and set aside the charges. The new motion was dismissed. The criminal charges against Gerald Phillips, Roger Parry and Curragh were allowed to proceed to trial.

In February 1995 the trial of Curragh, Gerald Phillips, and Roger Parry commenced in Pictou County, Nova Scotia. The trial lasted 44 days. Twenty-three witnesses were heard. Many of the days of the trial proceedings addressed the issue of the sufficiency of the disclosure by the Crown prosecutor. On March 2, 1995, the trial judge, Mr. Justice Robert Anderson, made a fateful telephone call to the prosecutor's office demanding that the lead prosecutor be taken off the case. The lead

---

[18] *R. v. Curragh Inc.* (1993), 124 N.S.R. (2d) 59 (N.S. Prov. Ct.).
[19] *Ibid.* at 66.

prosecutor learned of this telephone call and demanded a mistrial. Justice Anderson refused to grant a mistrial.[20]

On June 9, 1995, the three accused successfully brought a motion to stay the criminal prosecution against them on the basis of the Crown misconduct relating to failure to provide full, complete and timely disclosure. The legal right of an accused to full, complete and timely disclosure is discussed in chapter 6. Justice Anderson granted the defence motion to stay the charges on the basis that section 7 of the *Charter* had been violated as a result of the continuing non-disclosure and late disclosure of evidence to the defence by the prosecutor. Justice Anderson, in his reasons for the decision to stay all of the criminal charges, said:

> The applicants herein have basic rights and the authorities are clear that an accused is entitled to a fair criminal trial and has a right to make full answer and defence pursuant to the provisions of the Canadian *Charter of Rights and Freedoms* and the *Criminal Code* of Canada. I find that the lack of disclosure or non-disclosure, the timing of disclosure and the nature of the disclosure amounts to an infringement of the section 7 right to make full answer and defence and infringes on the right of a fair trial. Such a breach demands a remedy under section 24(1) of the *Charter*.[21]

The decision of Justice Anderson to stay all charges against Gerald Phillips, Roger Parry, and Curragh, was appealed to the Nova Scotia Court of Appeal. The argument before the court focused on the fateful telephone call by Mr. Justice Anderson to the prosecution office demanding that the lead prosecutor be taken off the case. On December 1, 1995, the Nova Scotia Court of Appeal overturned the stay of Justice Anderson and ordered a new trial. Writing for the court, the Chief Justice of Nova Scotia, Justice Hallett said:

> We are of the opinion the appeal should be allowed for the following reasons:
>
> (i)   in calling Mr. Herschorn and making the remarks to which we have been referred, the learned trial judge exhibited an appearance of bias that incurably infects his decision on the stay;
>
> (ii) the learned trial judge failed to make an inquiry and a proper determination whether evidence that had not been disclosed or was disclosed late was material to the respondents' ability to make full answer and defence as the trial judge was required to do...
>
> (iii) stays of proceedings are only granted in the clearest of cases.[22]

[20]   *R. v. Curragh Inc.*, (1995), 146 N.S.R. (2d) 163 (N.S.S.C.), rev'd (1995), 146 N.S.R. (2d) 161, aff'd (1997), 113 C.C.C. (3d) 481 (S.C.C.).
[21]   *Ibid.* at 179.
[22]   (1995), 146 N.S.R. (2d) 161 at 162 (C.A.).

The Nova Scotia Court of Appeal's decision was appealed to the Supreme Court of Canada. On March 20, 1997, the Supreme Court of Canada, in a 7-2 split decision, held that the decision of the Nova Scotia Court of Appeal was correct and that there should be a new trial. Justices La Forest and Cory wrote the majority judgment, with Chief Justice Lamer, Justice L'Heureux-Dube, Gonthier, and Iacobucci concurring. Justices McLachlin, now Chief Justice of Canada, and Justice Major strongly dissented and would have upheld the trial decision of Justice Anderson to stay all charges against the three accused.

In the majority reasons, the Supreme Court of Canada was critical of the conduct of the trial judge and said the following:

> However, on March 2, 1995, when the trial was well under way the judge again called the senior member of staff. To make such a call during the trial was, to say the least, unfortunate if not ill advised. It was sufficient in itself to raise the issue of apprehension of bias. Further, the words of the trial judge during this conversation confirmed that there was a reasonable apprehension of bias. He clearly expressed his displeasure with the manner in which the Crown attorney was conducting the case. The trial judge recommended that he be removed from the case and if he were not he would take steps "to secure that end". He thereby interfered with the Crown's conduct of its case, and so became inappropriately involved in the fray.[23]

The majority judgment went on to make further comments with respect to the importance of the independence and impartiality of a trial judge. The majority reasons stated:

> The right to a trial before an impartial judge is of fundamental importance to our system of justice. Should it be concluded by an appellate court that the words or actions of a trial judge have exhibited bias or demonstrated a reasonable apprehension of bias then a basic right has been breached and the exhibited bias renders the trial unfair. Generally the decision reached and the orders made in the course of a trial that is found by a court of appeal to be unfair as a result of bias are void and unenforceable.[24]

Although the Supreme Court of Canada released its decision ordering a new trial of Gerald Phillips, Roger Parry, and Curragh on March 20, 1997, no new trial was ever commenced. In the meantime, Curragh had been petitioned into bankruptcy. There was a thorough review of the prospect of proceeding with a new trial against Gerald Phillips and Roger Parry. On June 30, 1998, a public announcement was made that a second trial would not proceed.

---

[23] (1997), 113 C.C.C. (3d) 481 at 485 (S.C.C.).
[24] *Ibid.* at 486.

The legal response of prosecuting the corporate mine owner, Gerald Phillips and Roger Parry was never satisfactorily completed. The prosecutor's conduct and the decision of the trial judge both appear to have fallen below a reasonable standard of judicial conduct. The high level of disappointment and dissatisfaction with the judicial system is another part of the legacy of the Westray disaster. This dissatisfaction, not completing the first trial and then the decision to not proceed with a second trial, was referred to in the political and legislative debates regarding Bill C-45. The legal response to the Westray disaster was followed by the Inquiry. The Inquiry had a much more positive contribution than did the legal response to the Westray disaster.

THE PUBLIC INQUIRY

Less than a week after the explosion, Mr. Justice Peter Richard of the Nova Scotia Supreme Court was appointed to conduct an Inquiry into the disaster by Premier Donald Cameron. The Inquiry was given wide powers to investigate the causes of the Westray mine explosion. Justice Richard was asked to determine whether the mine had been properly established and operated, and whether any defects or neglects in the mine and its operation had contributed to the explosion. Justice Richard was asked to determine if the explosion could have been prevented and whether the mine's managers and owners had complied with all mining safety statutes and regulations. The mandate of the Inquiry included the investigation of all other matters that were considered relevant, leaving the door open for an investigation into the political and financial circumstances that may have contributed to the establishment and subsequent development of the mine.[25]

The United Steel Workers of America union became the certified bargaining agent for the workers at Westray shortly after the explosion. A union certification drive prior to the explosion had been unsuccessful. The United Steel Workers of America union (the "Union") participated in the Inquiry on behalf of the bargaining unit workers. The Union was highly critical of management at the Westray mine. The Union urged that Justice Richard, at the close of the Inquiry, consider the creation of a new criminal offence that would impose criminal liability on corporate executives, directors and other corporate agents for failing to ensure that the corporation they represented maintained an appropriate standard of

---

[25]   Westray Mine Public Inquiry, 1997 at 82.

health and safety in the workplace. The Union also recommended a creation of an offence of corporate killing in the *Criminal Code* of Canada. Further, the Union recommended that OHS legislation be broadened to identify potential liability of officers and directors of corporations.

Justice Richard heard many witnesses, interested parties and experts throughout the lengthy Inquiry. In his final report, Justice Richard posed six rhetorical "what if" questions forming the underpinnings of the Inquiry that, paraphrased, are as follows:[26]

(i)   What if Westray's CEO had sent a message to management that safety was paramount?

(ii)   What if the mine managers had conscientiously directed compliance with safety procedures?

(iii)   What if the *Coal Mines Regulation Act* had been applied and enforced by the Department of Labour?

(iv)   What if the Department of Natural Resources had fulfilled its legislative responsibility and ensured, before issuing permits, that Westray's plans were safe and being followed on a regular basis?

(v)   What if Westray miners had voted in favour of joining a union, as they had the chance to do in January 1992?

(vi)   What if a Department of Labour inspector had gone underground when he visited the mine on May 6, 1992 to ensure compliance with orders issued several days earlier?

The Inquiry report made a total of 74 recommendations, which ranged from the criminal liability of corporate executives and directors to misplaced work incentive plans based solely on bonuses for productivity. Other recommendations dealt with the ventilation of the mine, air quality, safety equipment and proper mining methods. Yet others focused on methane gas and the requirement of monitoring and measuring methane gas levels, as well as the operation of mining equipment in high-methane concentration areas. There were also recommendations dealing with the issue of coal dust and the necessity of controlling its spread with stone dust. Ground control issues in the mine were also the subject of recommendations. All of the Inquiry recommendations are included as Appendix "C" to this book.

---

[26]   *Ibid.* at 12-13.

Many of the recommendations from the Inquiry dealt with the issue of health and safety regulations. The need for a clear mandate and accountability of OHS regulators was also addressed. The Inquiry recommended legislative changes regarding the issuance of underground coal mining permits and the granting of approval of mine plans. As well, the recommendations dealt with emergency procedures, and processes to be followed during mine rescues.

The Inquiry recommendation that set the groundwork for Bill C-45 was Recommendation 73, which addressed the accountability of corporate executives and directors for ensuring the health and safety of workers was protected. Bill C-45 was a direct, albeit eventual, result of political action taken on the basis of Recommendation 73; it states:

> The Government of Canada, through the Department of Justice, should institute a study of the accountability of corporate executives and directors for the wrongful or negligent acts of the corporation and should introduce in the Parliament of Canada such amendments to legislation as are necessary to ensure that corporate executives and directors are held properly accountable for workplace safety.[27]

Years after it was made, Recommendation 73 was at the heart of the legislative and public policy debates, as legislators, unions, regulators, activists, and workplace stakeholders attempt to fashion an appropriate legislative response to the Westray disaster. This included attempts to strike a balance between workers' health and safety on the one hand, and requiring corporations to implement the internal responsibility necessary to self-regulate workplace health and safety on the other hand. The end of the Inquiry resulted in the beginning of a variety of legislative responses.

LEGISLATIVE RESPONSES

In Nova Scotia, the legislative response to the Westray disaster and the Inquiry was reasonably prompt. A new *Occupational Health and Safety Act* was introduced by the Nova Scotia government in 1996.[28] The new OHS legislation applied to the province itself,[29] every agency of the government of the province of Nova Scotia,[30] and it took precedence over any other general or provincial legislation of the province of Nova

---

[27]   *Ibid.* at 57.
[28]   S.N.S. 1996, c. 7, as am. by S.N.S. 2000, c. 28, ss. 86-87.
[29]   *Ibid.*, s. 4(1).
[30]   *Ibid.*, s. 4(2)(a).

Scotia, where any conflict existed.[31] A number of new legal duties and precautions were placed on employers,[32] contractors,[33] constructors,[34] suppliers,[35] employees,[36] self-employed persons,[37] owners,[38] persons who provide occupational health or safety services[39] and architects and professional engineers.[40] The new Nova Scotia *Occupational Health and Safety Act* enhanced the role of health and safety committees and provided the first legislated definition of the internal responsibility system in Canada.[41]

No specific duty or legal responsibility was added to the new Nova Scotia *Occupational Health and Safety Act* for directors and corporate executives. Although many of the Nova Scotia OHS legislative changes were modelled on Ontario's *Occupational Health and Safety Act*,[42] Nova Scotia did not follow Ontario's lead and establish legal duties, and corresponding liabilities, for directors and officers of corporations. Therefore, although the legislative response in Nova Scotia was prompt and progressive, it did not add any legal duties for directors, officers, and senior executives for OHS violations.

The federal government was slower to respond to the Westray disaster, the Inquiry and in particular Recommendation 73. In April 1999, the MP for Pictou-Antigonish-Guysborough, Peter MacKay, later elected as the leader of the federal Progressive Conservative Party, brought forward a Private Member's Motion 455, requesting that:

> The *Criminal Code* or other appropriate federal statutes should be amended in accordance with Recommendation 73 of the Province of Nova Scotia's Public Inquiry into the Westray disaster, specifically with the goal of ensuring that corporate executives and directors are held properly accountable for workplace safety.[43]

---

[31] *Ibid.*, s. 5.
[32] *Ibid.*, s. 13.
[33] *Ibid.*, s. 14.
[34] *Ibid.*, s. 15.
[35] *Ibid.*, s. 16.
[36] *Ibid.*, s. 17.
[37] *Ibid.*, s. 18.
[38] *Ibid.*, s. 19.
[39] *Ibid.*, s. 20.
[40] *Ibid.*, s. 21.
[41] See Chapter 5, The Internal Responsibility System.
[42] R.S.O. 1990, c. O.1.
[43] *House of Commons Debates*, 214 (23 April 1999) at 1215 (P. MacKay), online: <http://www.parl.gc.ca/36/1/parlbus/chambus/house/debates/214_1999-04-23/HAN214-E.HTM>.

Mr. McKay clearly linked his motion requesting amendments to the *Criminal Code* to Recommendation 73 from the Westray Inquiry. His speech introducing Motion 455 in the House of Commons included the following comments:

> The overriding attempt behind this motion would be to remind government and in fact all parliamentarians that we in this place and in provincial legislatures throughout the country must do everything in our power to ensure that there is a safe workplace for those who are engaged in labour activity. I speak not only of mines, I speak of fish plants. I speak of fishermen, of course, fisherpersons on the water. I speak of any factory and any situation where workers might find themselves encountering danger...I suggest quite strongly that knowing criminal sanctions or other disciplinary acts of retribution exist is the most direct way to ensure that those with the implicit responsibility for ensuring safety will abide. This would lead to a higher level of accountability among executives, CEOs and management in companies that directly benefit from what might be phrased expediency over safety in the workplace.[44]

On October 21, 1999, Alexa McDonough, MP for Halifax and then leader of the federal New Democratic Party, tabled Bill C-259.[45] This bill made corporations, directors and officers specifically accountable for workers' health and safety under the *Criminal Code*. This bill proposed OHS criminal liability for corporations, directors and officers with jail terms and fines in the millions of dollars. While the bill received some public interest and support, it died on the Order Paper after having only passed First Reading in October 2000, when a federal election was called.

Ms. McDonough's Private Members Bill C-259 was the first legislative response to the Westray disaster. Bill C-259 established four different scenarios in which a corporation would be guilty of an offence of which an individual could be found guilty. With some similarity to the eventual introduction of Bill C-45, Ms. McDonough's Private Members Bill purported to define when, by the actions of management of a corporation, the corporation became liable for criminal offence. Management of a corporation, in respect of an act of omission that amounted to a criminal offence, was defined in Bill C-259 as follows:

> (*a*) the one or more persons who, being directors or officers of the corporation are responsible for the direction and control of the part of the activity of the corporation in respect of which the act or omission occurs; and

---

[44] *Ibid.*

[45] *An Act to amend the Criminal Code (criminal liability of corporations, directors and officers)*, 2d Sess., 36th Parl., 1999 (1st reading 21 October 1999).

(*b*)   the one or more persons to whom the corporation has delegated the day-to-day management of that part of the activity of the corporation.[46]

Bill C-259 would have held a corporation guilty and subject to, on summary conviction, a fine not exceeding $50,000; or, on indictment, to a fine not exceeding $1 million, or, in the case of an offence of murder or manslaughter, not exceeding $10 million.[47]

Ms. McDonough's Bill also proposed a new offence for directors and officers of a corporation that were found guilty of one of the new offences established by the bill. If a director or officer of the corporation alone, or with others, authorized the act or omission that constituted an offence by the corporation, or knew or ought to have known that the act or omission constituted an offence, would be guilty of an offence themselves.[48] If convicted, a director or officer of a corporation would be liable on conviction to a fine of not more than $10,000 for each day on which the unsafe working condition was shown to have existed while the individual held the office of officer or director, or an imprisonment for not more than seven years, if the unsafe working conditions did not result in the death of a worker, and, imprisonment for life if the unsafe working conditions resulted in the death of any person.[49]

In February 2001, a second Private Members Bill, C-284,[50] dealing with similar subject matter to Bill C-259, was introduced by Bev Desjarlais, New Democratic Party member of parliament for Churchill, Manitoba, and received First Reading. The purpose of the bill was to ensure that "where a member of the staff of a corporation commits an offence by an act or omission on behalf of the corporation, the corporation, its directors and officers may, in certain circumstances also be guilty of the offence."[51]

Bill C-284 appeared to gain more interest and support in the House of Commons than Bill C-259, even though it was very similar in content. The introduction of Bill C-284 marked the beginning of a move towards legislative change in Canadian criminal law, and away from the directing mind theory of corporate liability. Traditionally, for a corporation to be held liable for a criminal offence, prosecutors must demonstrate that the person responsible was the directing mind of the corporation and "that he

---

[46]   *Ibid.*, s. 1, amending s. 467.3(1)
[47]   *Ibid.*, s. 1, amending s. 467.3(3)
[48]   *Ibid.*, s. 1, amending s. 467.4.
[49]   *Ibid.*, s. 1, amending s. 467.6(1).
[50]   *An Act to amend the Criminal Code (offences by corporations, directors and officers)*, 1st Sess., 37th Parl., 2001 (1st reading 26 February 2001).
[51]   *Ibid.*, summary.

or she had knowledge of all the facts which caused the harm, and acted or failed to act in contravention of the law."[52]

Bill C-284 offered the same definition of management of the corporation to establish criminal liability as did Bill C-259. Bill C-259 had the same fines proposed for corporations convicted as Ms. McDonough's Private Members Bill. Also, like Bill C-274, Bill C-284 proposed that directors and officers could face up to life imprisonment if their criminal act or omission resulted in the death of any person.

Bill C-284 also provided that if it were shown that the corporation's staff committed the act or omission, the burden shifted to the corporation to show that action or omission was unauthorized and not tolerated by the corporation. This reverse onus was a controversial provision of the bill. Bill C-284 was introduced and was debated in the House of Commons during Second Reading on September 20, 2001 and November 8, 2001, respectively. In February of 2002, the House of Commons referred the bill to the Standing Committee on Justice and Human Rights. While Bill C-284 was approved in principle by all parties in the House of Commons, the reverse onus in the bill also raised concerns regarding constitutional and *Charter* issues,[53] and it was eventually withdrawn from House business. The subject matter of the bill was submitted to the House of Commons Standing Committee on Justice and Human Rights. This was the next step on the road to Bill C-45.

STANDING COMMITTEE ON JUSTICE AND HUMAN RIGHTS

The House of Commons of the Parliament of Canada has a number of standing committees of which members of parliament are members. These committees are constituted for the purpose of doing work to assist Parliament. One of the most important committees is the Standing Committee on Justice and Human Rights (the "Committee"). The Committee, among other things, reviews new legislation with respect to criminal justice and human rights. It was to this committee that Bill C-284 was referred for consideration in February of 2002.

While Bill C-284 was eventually withdrawn, the Committee then heard evidence from witnesses and various interested parties on the issue

---

[52] Lawrence McBrearty, *Brief to Standing Committee on Justice and Human Rights on Amendments to the Criminal Code*, (Paper presented to the United Steelworkers of America, 8 May 2002), at 10, online: <http://www.uswa.ca/eng/westray/westray_brief.PDF>.

[53] *House of Commons Debates,*119 (15 September 2003) at 550 (Vic Toews), online: <http://www.parl.gc.ca/37/2/parlbus/chambus/house/debates/119_2003-09-15/HAN119-E.HTM>

of corporate criminal liability. While the Westray disaster was discussed in detail, the Committee stressed the necessity of exploring the larger issue of the need for corporations to be more accountable. There was a growing consensus at the Committee of the need to reform the law of criminal liability of corporations.

Over the course of a few weeks in May 2002, the Committee heard evidence from numerous stakeholders and interested parties as to the need for and feasibility of introducing legislation addressing corporate criminal liability. The witnesses ranged from Westray victims' families to union members; members of the legal profession to various advocacy groups. One of the discussions at the Committee was that while OHS legislation gives workers the right to refuse work that is hazardous or dangerous, "in reality, this very important right is not enforceable because workplace harassment and intimidation by managers effectively prevent workers from enforcing their rights."[54] The argument was made that although workers may exercise some degree of control over their own OHS, they are often reluctant to raise concerns due to concerns about job security.

In their presentation to the Committee, the Union encapsulated the four principles that any new legislation on corporate criminal liability should attempt to address:[55]

(i)   Officers and directors of corporations have a duty to protect their employees' lives at work.
(ii)   Officers and directors may be held criminally liable for acts or failure to act because of wilful blindness and criminal negligence when they fail to carry out their duties and serious injury or illness results for their employees.
(iii)   Corporations may be held criminally liable when their management operates the company in a fashion that ignores or is wilfully blind to the health and safety of their employees and death or serious injury or illness results for the employees.
(iv)   Officers and directors and the corporation may be prosecuted and convicted of separate offences arising out of the same circumstances.

---

[54]   Maria York, "To the Standing Committee on Justice and Human Rights", Submission by the Canadian Council for Rights of Injured Workers (2 May 2002), online: www.ccriw.com/scjhr.pdf. See also the comments of Peter MacKay in the Minutes of Proceedings, at 1025, online: <www.parl.gc.ca/InfoComDoc/37/1/JUST/Meetings/Evidence/JUSTEV87-E.HTM>, [*Minutes of Proceedings*].
[55]   *Minutes of Proceedings, ibid.* at 1540.

Evidence was also presented to the Committee on the status of workplace health and safety regimes across Canada. Even though organized labour participates in health and safety committees across Canada, the internal responsibility system came under scrutiny and criticism. Strong criticisms of OHS legislation, the internal responsibility system, and its enforcement by government regulators was made by organized labour. Organized labour views were expressed, in part, as follows:

> The [occupational health and safety] system began to unravel in the 1990s with de-regulation and the systematic attempt by businesses to get government off their backs. The regulations stagnated, with new hazards unacknowledged or unrecognized in the regulations. The existing regulations were subjected to weaker standards or types of standard setting. The chief form of deregulation was simply to cease to enforce the law. The enforcement branch was radically weakened through some euphemisms, such as rationalization, and the inspectors were persuaded or intimidated into pursuing the policy of voluntary compliance, meaning that employers should conform to the law only if they felt like it.[56]

While the issues being debated remained emotionally charged, there was a recognition that change in the criminal liability of corporations was needed. Various models of liability were surveyed and it was concluded that no one current model was wholly adequate.

The Committee's singular recommendation, presented on June 6, 2002, was "that the Government table in the House legislation to deal with the criminal liability of corporations, directors, and officers."[57] In its Report, the Committee considered Westray Inquiry Recommendation 73 and the consensus on the need to reform corporate criminal liability in Canada. There was no reference in the Committee's recommendation to extending OHS criminal liability to all organizational decision-makers who directed how work was to be performed.

In response to the Committee's Report, the federal government issued a Response Paper in November 2002. The government outlined the need to develop a made-in-Canada approach to criminal law reform and health and safety enforcement. That approach would continue to recognize existing OHS legislative regimes, such as amendments to Part II of the *Canada Labour Code*.[58] There would also be a continued emphasis on the internal responsibility system.[59] All Canadian OHS legislation has

---

[56] *Ibid.* at 1130.
[57] House of Commons, Standing Committee on Justice and Human Rights; online: <www.parl.gc.ca/InfoComDoc/37/1/JUST/Studies/Reports/JustRP15-E.htm>.
[58] S.C. 2000, c. 20.
[59] See Chapter 5, The Internal Responsibility System.

established the right to know about hazards in the workplace, the right to participate in correcting these hazards, and the right to refuse dangerous work for workers. OHS statutes and offences would still play an important role in protecting worker and public safety. The key to the government's model was that it would have criminal law sanctions applying "to all persons without regard to how they choose to organize their affairs."[60] The government's response was, however, a departure from Recommendation 73 of the Westray Inquiry. The key, according to the government position, was that:

> Officers and directors should not be singled out and have liability imposed on them either generally or with respect to safety simply because of the way the business is structured. They should be held criminally responsible for the way they carry out their responsibilities and not be subject to criminal liability in the absence of personal fault simply because of their position in the corporation.[61]

INTRODUCTION OF BILL C-45

The federal government was slow in developing its response to the Westray disaster. The Inquiry and the growing consensus of the need for criminal law reform with respect to corporations influenced the government's response. Although legislative action was promised in November 2002, no legislation was forthcoming for months. In March 2003, growing impatience with the government's inaction resulted in Alexa McDonough introducing Bill C-418, *An Act to amend the Criminal Code (criminal liability of corporations, directors and officers).*[62] This bill dealt with the same material as the previous two NDP Private Members bills of the same name. The NDP kept the subject of reforming the *Criminal Code* regarding the criminal liability of corporations in the forefront of the federal political arena.

Rather than choosing to adopt Ms. McDonough's Private Members Bill, however, the federal government finally provided its own response. On June 12, 2003, the federal government tabled the long-awaited new legislation, Bill C-45, *An Act to amend the Criminal Code*

---

[60]   Canada, Department of Justice, *Government Response to the Fifteenth Report of the Standing Committee on Justice and Human Rights* (Ottawa: Department of Justice, 2002) at 9, online: <http://canada.justice.gc.ca/en/dept/pub/ccl_rpm/>.

[61]   *Ibid.* at 13.

[62]   2nd Sess., 37th Parl., 2003 (1st reading 20 March 2003).

(criminal liability of organizations).[63] Some of the highlights of Bill C-45 include:[64]

(i)   Corporate criminal liability is no longer dependent on the actions of a directing mind, or senior officer, of the corporation;

(ii)   Liability may be engaged even in instances where the criminal elements of *mens rea* and *actus reus* may not be derived from the same individual;

(iii)   All members of an organization, from directors to officers to employees, agents and contractors may be found criminally liable;

(iv)   Negligence may be found in cases where *mens rea* is aggregated amongst numerous individuals;

Bill C-45 amended the *Criminal Code* such that crimes of intent or recklessness may be attributable to the organization, even where a senior officer is not party to the offence, if the senior officer had knowledge of the commission of the offence and failed to take all reasonable steps to prevent the occurrence. The introduction of the new term "senior officer" was pivotal to the changes to corporate criminal liability.

---

[63]   2nd Sess. 37th Parl., 2003 (1st reading 12 June 2003). The bill received Royal Assent on 7 November 2003 as 2003, c. 21.

[64]   Bill C-45, Legislative Summary. See also the speech of Mr. Richard Marceau, Bloq Québecois MP for Charlesbourg—Jacques-Cartier, *supra* note 14 at 1605, where he outlines the 8 key points of the legislation:

1. [T]he use of the term "organization", rather than "corporation" … [which] will broaden the definition, thereby affecting more institutions.
2. [C]ompanies can now be held criminally liable for the acts of their employees who are not necessarily in positions of authority.
3. [T]he act of committing a crime … and the intent to commit a crime … of criminal offences attributed to companies and other organizations no longer need be the work of the same person.
4. [T]he category of persons whose acts or omissions can constitute the … criminal act … is broadened to include all employees, representatives or contractors.
5. [W]ith regard to … criminal negligence, the fault can now be attributed to the organization to the extent that one of the senior officers of the organization can be charged with the offence.
6. In the case of deliberate crimes, an organization can now be held responsible for the actions of its senior officers to the extent that a senior officer is party to the offence, directs other employees to commit an offence or, knowing that an offence will be committed by other employees, does nothing to prevent it.
7. [P]lace the onus explicitly on anyone who undertakes to direct the work of other employees to take all reasonable steps to prevent bodily harm to these employees.
8. [T]he bill also contains provisions for establishing general sentencing principles. and probation conditions in respect of the organizations.

One of the most important aspects of the new legislation is that the word "corporation" has been amended and replaced with the new term "organization", in the *Criminal Code*. This allows a broader spectrum of associations, groups and bodies to be captured by the new OHS duty and existing criminal legislation. This aspect of Bill C-45 will be discussed in more detail in chapter 4 of this book.

In Bill C-45, the government stressed that wanton or reckless disregard for the lives of persons in the workplace and breaching the new legal duty to take reasonable steps in avoiding foreseeable harm to the person or the public could result in a new offence of OHS criminal negligence. The new standard of fault for corporate liability would apply to all offences committed by an organization, including those involving workplace safety. Essentially, any time a senior officer with policy-making or operational authority does not take liable action to prevent representatives of the organization from committing an offence, the organization may be held criminally liable.

In the Bill C-45 debate, the federal government responded to many of the issues raised by the Committee Report.[65] It emphasized the reactive nature of criminal law and recognized that OHS legislation, on the other hand, proactively sets out prevention requirements. The result was that the federal government stated that it intends to continue to rely on the OHS regulatory laws outside of the criminal arena in addition to the Bill C-45 amendments. The point was brought home with the following excerpt from the government's response to the Committee's report: "The criminal law must be reserved for the most serious offences, those that involve grave moral faults... the Government does not intend to use the federal criminal law power to supplant or interfere with the provincial regulatory role in workplace health and safety."[66]

Before Bill C-45, corporate criminal liability was established when proof of criminal activity based on the identification theory was met. Under this legal doctrine, the corporate employee who committed the criminal offence must be the centre of the corporate personality; a vital organ of the body corporate, alter ego of the corporation or its directing mind.[67] Generally, this theory of liability tended to limit responsibility to those in the upper echelons of senior management. Bill C-45 has

---

[65] Government Response to Fifteenth Report of the Standing Committee on Justice and Human Rights (Ottawa: Department of Justice, 2002) at 9, online: <http://canada.justice.gc.ca/en/dept/pub/ccl_rpm/>.

[66] *Ibid.* at 2.

[67] *R. v. Canadian Dredge & Dock Co.*, [1985] 1 S.C.R. 662 at 682. The identification doctrine is discussed in more detail below, Section 2(g).

broadened this concept of corporate criminal liability by extending the category of individuals whose actions might engage corporate responsibility. This was in the form of a new defined term, "representative". This term includes directors, partners, employees, members, agents and contractors of the organization. Bill C-45 was also designed to add an OHS legal duty to the *Criminal Code* that would dovetail with the existing crime of criminal negligence and the new criminal liability of an organization. This new OHS legal duty, when breached, will lead to a charge of the new offence of OHS criminal negligence.

Legal and policy criticisms of the narrow and simplistic nature of the identification theory were widespread.[68] Bill C-45 was intended to address that problem. A review of the general law relating to criminal negligence and organization liability is covered in chapter 3. That chapter will also review some previous failed attempts to prosecute corporations for criminal negligence for workplace injury and death prior to Bill C-45. What follows is the legislative development and history of Bill C-45 itself.

BILL C-45 LEGISLATIVE HISTORY

House of Commons		Senate	
1st Reading	June 12, 2003	**1st Reading**	October 27, 2003
Debate(s) at 2nd Reading	September 15, 2003; September 19, 2003	**Debate(s) at 2nd Reading**	October 29, 2003
2nd Reading	September 19, 2003	**2nd Reading**	October 29, 2003
Committee	Justice and Human Rights	**Committee**	Legal and Constitutional Affairs
Committee Meeting(s)	October 22, 2003	**Committee Meeting(s)**	October 30, 2003
Committee Report	October 23, 2003	**Committee Report**	October 30, 2003

---

[68]   See generally, *R. v. Canadian Dredge & Dock Co.* [1985] 1 S.C.R. 662.

House of Commons		Senate	
Debate(s) at Report Stage		**Debate(s) at Report Stage**	
Report Stage	October 27, 2003	**Report Stage**	
Debate(s) at 3rd Reading	October 27, 2003	**Debate(s) at 3rd Reading**	October 30, 2003
3rd Reading	October 27, 2003	**3rd Reading**	October 30, 2003

Royal Assent: November 7, 2003
Statutes of Canada: 2003, c.21

CHAPTER 3

THE LAW OF CRIMINAL NEGLIGENCE

THE OFFENCE OF CRIMINAL NEGLIGENCE

Bill C-45 has established a new legal duty for the health and safety of workers and the public in the *Criminal Code*. If breached, this duty then gives rise to the new offence of OHS criminal negligence. This new duty dovetails with the existing provisions of the *Criminal Code* dealing with criminal negligence. Therefore, an introduction to the law of criminal negligence is vital for a comprehensive understanding of Bill C-45 and its implications.

Criminal negligence is an offence under the *Criminal Code* that can arise from either acts or omissions of the accused. If the accused was under a legal duty and breached that duty by an act or omission and did so with wanton or reckless disregard for the lives or safety of other persons, this amounts to the offence of criminal negligence. There are a number of legal duties imposed under the *Criminal Code* that would form part of the basis of the offence of criminal negligence. They include a duty to dispense rioters,[1] a duty of care with respect to the handling of explosives,[2] duty of care regarding the handling of firearms,[3] a duty to provide the necessaries of life for a child under the age of sixteen, a spouse or common-law partner, and a person under his charge if that person is unable to provide himself with the necessaries of life,[4] a duty of persons undertaking acts dangerous to life,[5] a duty to every one who undertakes to administer surgical or medical treatment to another

---

[1] R.S.C. 1985, c. C-46, s. 33.
[2] *Ibid.*, s. 79.
[3] *Ibid.*, s. 86.
[4] *Ibid.*, s. 216.
[5] *Ibid.*, s. 217.

person,[6] a duty to safeguard openings in ice and excavations[7] and other duties relating to non-*Criminal-Code* duties such as *Highway Traffic Act* and OHS regulatory duties. However, the mere breach of a duty imposed by a provincial or federal statute, other than the *Criminal Code*, does not *per se* constitute criminal negligence.[8]

To establish the crime of criminal negligence, the prosecution is not required to show proof of intention or deliberation, but must show wanton or reckless disregard for the lives or safety of others. A large number of criminal negligence cases relate to motor vehicle accidents and related injury and death. For example, an accused may be convicted of criminal negligence on proof of driving that amounts to a marked and substantial departure from the standard of a reasonable motor vehicle driver in circumstances where the accused either recognized then disregarded an obvious and serious risk to the lives and safety of others or, alternatively, gave absolutely no thought to that risk whatsoever.[9]

The law of criminal negligence in Canada is found in sections 219, 220 and 221 of the *Criminal Code*[10] which provide as follows:

219. (1) Every one is criminally negligent who

(*a*)   in doing anything, or

(*b*)   in omitting to do anything that it is his duty to do,

shows wanton or reckless disregard for the lives or safety of other persons.

(2) For the purpose of this section, "duty" means a duty imposed by law.

220. Every person who by criminal negligence causes death to another person is guilty of an indictable offence and liable

(*a*)   where a firearm is used in the commission of the offence, to imprisonment for life and to a minimum punishment of imprisonment for a term of four years; and

(*b*)   in any other case, to imprisonment for life.

221. Every one who by criminal negligence causes bodily harm to another person is guilty of an indictable offence and liable to imprisonment for a term not exceeding ten years.

---

[6]   *Ibid.*, s. 216.
[7]   *Ibid.*, s. 263(1), (2).
[8]   *R. v. Titchner* (1961), 131 C.C.C. 64 (Ont. C.A.)
[9]   *R. v. Sharpe* (1984), 12 C.C.C. (3d) 428 (Ont. C.A.).
[10]   R.S.C. 1985, c. C-46.

The *Criminal Code* provides a definition of bodily harm. This definition has not been amended by Bill C-45. Bodily harm means, "...any hurt or injury to a person that interferes with the health or comfort of the person and that is more than merely transient or trifling in nature".[11] The legal interpretation and jurisprudence surrounding the phrase "bodily harm" is quite limited. Mr. Justice Borins typified most judicial commentary on the meaning to be given to the phrase "bodily harm", when he said, "bodily harm really requires no explanation, that the words should be given their ordinary meaning and that the harm need not be really serious..."[12] The Alberta Court of Appeal has stated that bodily harm, "...includes any hurt or injury which interferes with the health or comfort; it need not be permanent but must be more than merely transient or trifling."[13]

In a somewhat more helpful definition of bodily harm, the British Columbia Court of Appeal in *R. v. Dickson*, said the following:

> The element of interference with comfort, which is all that the definition requires, must have continued for some time after that. The interference with comfort resulted from a significant injury – one which cannot be described as trifling. There is no necessary connection at all between the duration of the injury and the question whether it is trifling – a life-threatening injury is often resolved in a short time. Transient does not relate to time but, in this context, it is simply insupportable to describe as transient an injury that "lasts no longer than a month.[14]

English courts have also considered the meaning of bodily harm, since the nature of criminal negligence in the United Kingdom is virtually identical to the Canadian offence. The English Court of Appeal has held that bodily harm includes injury to the nervous system and brain, therefore it could include an injury to those parts of the body responsible for mental and emotional faculties.[15] Further, the Supreme Court of Canada has held that psychological injury may also be characterized as bodily harm.[16] The limited jurisprudence with respect to bodily harm raises interesting questions with respect to the application of Bill C-45 and the new legal duty under section 217.1 of the *Criminal Code*.

For example, could failure of an employer to take reasonable steps to ensure than an ergonomically correct work station is available for office

---

[11] *Ibid.*, s. 2.
[12] *R. v. McNamara* (1979), 12 C.R. (3d) 210, at 225, Borins J.
[13] *R. .v. Martineau* (1988), 43 C.C.C. (3d) 417, at p. 427, Laycraft, J. aff'd [1990] 2 S.C.R. 633.
[14] *R.v. Dixon* (1988), 42 C.C.C. (3d) 318 at p. 332, Esson, J.A.
[15] *R.v. Chan-Fook*, [1994] 1 W.L.R. 689 at 695.
[16] *R. v. McCraw* (1991), 128 N.R. 299 (S.C.C.).

employees, which results in back problems or a repetitive strain injury, be the subject of a prosecution for OHS criminal negligence? In such an inquiry, the court would undoubtedly have to address the question of whether or not back strain or repetitive strain injury from poorly designed work stations would constitute bodily harm. Similarly, could a case be made out that an employer that fails to prohibit outright smoking in the workplace, in spite of an absence of any statutory or bylaw requirement, would be failing to take reasonable steps to prevent bodily harm if there is a body of scientific evidence that second-hand smoke in the workplace may cause respiratory, non-cancerous, and cancerous diseases?

The law of criminal negligence is currently in a state of some uncertainty in Canadian law. The law has generally accepted that there is no distinction between acts or omissions[17] and there is clearly no need to prove deliberateness and wilfulness. However, the fault element of the offence has been problematic. In the words of Justice Sopinka in *Anderson v. R.*,[18] "this area of the law, both here and in other common-law countries, has proved to be one of the most difficult and uncertain in the whole of the criminal field."[19] The difficulty arises from the *mens rea* or fault element; whether it is objective, subjective or to some extent both. In order to illustrate how the courts have approached this issue, it is best to consider the key elements of criminal negligence in relation to the current case law.

ACTS AND OMISSIONS

There has been some judicial and academic debate whether there is, or should be, a distinction between crimes of criminal negligence arising out of acts and those based on omissions. Although this distinction is not supported by the current case law, the Supreme Court of Canada has not completely closed the door on the potential for a distinction. This is primarily due to the fact that the law of criminal negligence, as mentioned before, is in a state of flux and the potential for development and changes remains.

In Ontario, the Court of Appeal in *R. v. Canhoto*[20] has held that the fault element for both acts and omissions is the same, measured by a

---

[17] See Chapter 3, (Acts and Omissions).
[18] (1990), 53 C.C.C. (3d) 481 (S.C.C.).
[19] *Ibid.* at 485.
[20] (1999), 140 C.C.C. (3d) 321 (Ont. C.A.).

uniform objective standard. In that case, the accused was charged with manslaughter on the basis of criminally negligent conduct contributing to the death of her two-year-old daughter. The child died from an inability to breathe and the aspiration of water which occurred while her grandmother forced water down the child's throat as part of a ritual to expel evil spirits from the child. The child resisted, kicking strenuously and screaming. The trial judge found that the accused was present during some of the forced-feeding water to the child, provided water and held the child's feet to restrain her. The trial judge also held that the accused knew or ought to have known that what the grandmother was doing to the child was capable of putting the child's life or safety at risk. In dismissing the appeal from the accused's conviction, the legality of both acts and omissions was considered to determine the appropriate standard for criminal negligence. In *Canhoto* Justice Doherty, speaking for the court, held that:

> The determination of fault based on a failure to direct one's mind to a risk can be applied equally to acts and omissions. The need for a uniform standard for the determination of criminal culpability is as important where the law imposes a duty to act and no action is taken, as it is in cases where a person engages in conduct which creates that same risk. It would run contrary to the principle of uniformity and the values underlying that principle if the criminal law were to distinguish between a parent who chooses not to administer a life-saving drug to his child, thereby risking the life of that child, and a parent who actually removes the needle containing the drug from the arm of the child, thereby creating the very same risk.[21]

Doherty J.A. went on to further explain:

> Section 219 defines criminal negligence in terms of actions (section 219(1)(a)) and omissions where there is a duty to act (section 219(1)(b)). It provides that either actions or omissions constitute criminal negligence where they show a wanton or reckless disregard for the lives or safety of others. The language of the *Criminal Code*, the primary source to be examined when determining the elements of a crime, suggests no difference between the fault element of criminally negligent conduct and the fault element of a criminally negligent failure to act where there is a duty to act.[22]

As a result of the decision in *Canhoto*, there is currently no legal distinction to be made between acts and omission in the context of criminal negligence. However, in either situation, the act or omission is in relation to the lives or safety of other persons. This part of the law with respect to criminal negligence will not change as a result of Bill C-45.

---

[21]   *Ibid.* at 330.
[22]   *Ibid.*

OBJECTIVE VS. SUBJECTIVE STANDARD

The most controversial and perhaps most difficult part of the offence of criminal negligence is in the standard to be applied to the fault element or *mens rea*. As was previously mentioned, the question of whether or not the standard of *mens rea* applicable to criminal negligence is objective, subjective or both has been reviewed extensively in the courts. Justice Sopinka in *Anderson,*[23] helpfully summarized the basis of this disagreement as follows:

> The use of the word "negligence" suggests that the impugned conduct must depart from a standard objectively determined. On the other hand, the use of the words "wanton and reckless disregard" suggests that an ingredient of the offence includes a state of mind or some moral quality to the conduct which attracts the sanctions of the criminal law. The section makes it clear that the conclusion that there is a wanton or reckless disregard is to be drawn from the conduct which falls below the standard. The major disagreement in the cases centres around the manner in which this conclusion is to be drawn.

> On the one hand, there are the cases that hold that it is to be done on an objective basis. If the conduct is a marked departure from the norm, then, based on the standard of an ordinary prudent individual, the accused ought to have known that his actions could endanger the lives or safety of others. On the other hand, there are cases that apply a subjective standard and require some degree of advertence to the risk to be proved. This may be done by inferring advertence from the nature of the conduct in the context of the surrounding circumstances. A refinement on the latter view is that a marked departure constitutes a prima facie case of negligence. The trier of fact may but is not obliged to infer the necessary mental element from the conduct which is found to depart substantially from the norm.[24]

The Supreme Court of Canada has considered this issue on two occasions and despite the opportunity to do so, has not provided conclusive direction on the issue of the standard to be applied.

The first time that the Supreme Court considered this issue they divided equally on the question of whether the fault element of criminal negligence required an objective or subjective standard as it applied to acts of commission[25] and omission[26]. The cases *R. v. Tutton*[27] and *R. v. Waite*[28], decided on the same day, show the complexity and different

---

[23] (1990), 53 C.C.C. (3d) 481 (S.C.C.).
[24] *Ibid.* at 485.
[25] *R. v. Waite,* [1989] 1 S.C.R. 1436. This case concerned criminal negligence resulting from a drinking and driving offence.
[26] *R. v. Tutton,* [1989] 1 S.C.R. 1392. This case involved criminal negligence resulting from the failure of parents to provide medical treatment to a child.
[27] *Ibid.*
[28] [1989] 1 S.C.R. 1436.

analysis available in this area of law. Justice McIntyre in *Tutton*, with whom Justices L'Heureux-Dubé and Lamer concurred, was of the view that an objective standard ought to be adopted. On the other hand, Madame Justice Wilson, with whom Chief Justice Dickson and Justice La Forest concurred, preferred a subjective standard.

Justice McIntyre, in *Tutton*, in adopting an objective standard, distinguished between an ordinary criminal offence that requires proof of a subjective state of mind and the offence of criminal negligence. McIntyre J. explained that:

> In choosing the test to be applied in assessing conduct under section 202 of the Code, it must be observed at once that what is made criminal is negligence. Negligence connotes the opposite of thought-directed action. In other words, its existence precludes the element of positive intent to achieve a given result. This leads to the conclusion that what is sought to be restrained by punishment under section 202 of the Code is conduct, and its results. What is punished, in other words, is not the state of mind but the consequences of mindless action.[29]

Justice McIntyre concluded that the test for criminal negligence was that of reasonableness and that "proof of conduct which reveals a marked and significant departure from the standard which could be expected of a reasonably prudent person in the circumstances will justify a conviction of criminal negligence."[30]

Justice Lamer concurred with the reasons of Justice McIntyre in *Tutton*. However, Justice Lamer was of the view that when applying the objective standard a "generous allowance for factors which are particular to the accused, such as youth, mental development, education"[31] should be considered. By considering these factors Justice Lamer felt that, regardless of whether a subjective or objective test is used, the result would be essentially the same.

Madame Justice Wilson expressed her concerns in *Tutton* that the strict objective standard adopted by Justice McIntyre may in fact create an absolute liability offence.[32] In light of the serious nature of the offence of criminal negligence and the potential criminal punishment, Justice Wilson concluded that:

> The phrase "wanton or reckless disregard for the lives or safety of other persons" signifies more than gross negligence in the objective sense. It requires some degree of awareness or advertence to the threat to the lives or safety of others or al-

---

[29]   [1989] 1 S.C.R. 1392 at para. 43.
[30]   *Ibid.*
[31]   *Ibid.* at para. 49.
[32]   *Ibid.* at para. 5.

ternatively a wilful blindness to that threat which is culpable in light of the gravity of the risk that is prohibited.[33]

In coming to this conclusion, Justice Wilson argued that proof of negligent conduct created an evidential burden on the accused to "explain why the normal inference of conscious awareness or wilful blindness should not be drawn."[34] Justice Wilson, therefore, would require a minimal intent requirement of awareness or advertence or wilful blindness to the prohibited risk.

The Supreme Court had another opportunity to consider this issue in *R. v. Anderson,*[35] which concerned a driving offence. In this case, Justice Sopinka attempted to synthesize the two tests for the fault element relied on in *Tutton* and *Waite*. In his judgment, Justice Sopinka suggested that what is important is not the test applied, whether it is objective or subjective, but rather the finding that the conduct is a marked departure from the standard.[36] Without that finding there is no need to apply either a subjective or objective standard as the conduct would not come within the parameters of the offence which requires such a conclusion. As Justice Sopinka explained, "the unfortunate fact that a person was killed added nothing to the conduct of the appellant...If driving and drinking and running a red light was not a marked departure from the standard, it did not become so because a collision occurred."[37]

Since the law in this area has not been finally settled by the Supreme Court, lower courts have been left somewhat to fend for themselves when applying the correct legal test for criminal negligence. In three Court of Appeal for Ontario cases, *R. v. Nelson,*[38] *R. v. Cabral*[39] and *R. v. Gingrich,*[40] that court has now consistently adopted the objective standard when considering the offence of criminal negligence, whether it be in relation to a commission of an act or a failure to act. The court in *Nelson* specifically stated:

---

[33] *Ibid.* at para. 14.
[34] *Ibid.* at para. 15.
[35] (1990), 53 C.C.C. (3d) 481 (S.C.C.).
[36] *Ibid.* at para. 486.
[37] *Ibid.* at para. 488.
[38] (1990), 54 C.C.C. (3d) 285 (Ont. C.A.).
[39] (1990), 54 C.C.C. (3d) 317 (Ont. C.A.).
[40] (1991), 65 C.C.C. (3d) 188 (Ont. C.A.).

The decisions of the Supreme Court of Canada in *Waite* and *Tutton* do not stand for the proposition that the objective test as enunciated by this court in *Waite* and *Sharp* is incorrect. Until the Supreme Court of Canada holds that to be the case, this court ought not to depart from the objective standard.[41]

The Court of Appeal for Ontario now applies the arguable lower, objective standard in the offence of criminal negligence. Justice Finlayson adopted this reasoning in *Gingrich* and concluded that, "the crime of criminal negligence is negligence. There is no need to import the concept of a subjective intent in order to obtain a conviction...I can think of no reason why this court should depart from what it has said in *Cabral* and *Nelson*."[42]

The fault element in criminal negligence, therefore, is primarily an objective standard. However, to be convicted of the crime of criminal negligence, there must be proof beyond a reasonable doubt that the accused showed wanton or reckless disregard for the lives or safety of other persons. This standard will continue under the Bill C-45 amendments and has not changed. Wanton or reckless disregard has not been defined in the *Criminal Code*. Therefore, the next section of this chapter will deal with case law considering and interpreting the meaning of wanton or reckless disregard, the important fault element of the offence of criminal negligence.

WANTON OR RECKLESS DISREGARD

Criminal negligence may be based on an act or omission under section 219 of the *Criminal Code*. In either respect, the fault element requires proof that the accused showed wanton or reckless disregard for the lives or safety of other persons. This fault element has not been amended under Bill C-45.

Simply put, conduct that is reasonable cannot be wanton.[43] That being said, the courts have taken an expansive approach in trying to define what is wanton and what is reckless. One of the more extensive considerations of the meaning of wanton or reckless disregard is found in Justice Cory's judgment in *R. v. Waite*.[44] That case involved a person who drove a car into a group participating at a Bethel Mennonite Church hayride resulting in the death of four and serious injury to the fifth

[41] (1990), 54 C.C.C. (3d) 285 at 289.
[42] (1991), 65 C.C.C. (3d) 188 at 199.
[43] *R. v. Morrisey,* [2000] 2 S.C.R. 90.
[44] (1986), 28 C.C.C. (3d) 326 at 342 (Ont. C.A.).

person. The driver of the car was repeated to have said: "Let's play chicken, see how close we can come to the group at the hayride." Following the accident, he went into the trunk of his car and threw one of the coolers of beer into a field in an apparent attempt to hide it. In this judgment, which was affirmed by the Supreme Court of Canada, Justice Cory attempted to synthesize the phrase "wanton or reckless disregard", by stating that:

> ...it should be noted that caution should be exercised in applying English cases dealing with what constitutes "recklessly" in different statutes in light of the precise wording of section 202. The section provides that everyone is guilty of criminal negligence who "shows" a wanton or reckless disregard for the lives and safety of others. The use of the word "shows" indicates that an objective test should be applied. The very driving or act of commission complained of should "show" or demonstrate the criminal negligence. The word "wanton" means "heedlessly". "Wanton" coupled as it is with the word "reckless", must mean heedless of the consequences or without regard for the consequences. If this is correct, then it is immaterial whether an accused subjectively considered the risks involved in his conduct as the section itself may render culpable an act done which shows a wanton or reckless disregard of consequences. The section itself indicates that from the very manner of driving or from the very act committed the *mens rea* of the accused can be inferred.[45]

In a more recent decision of the Superior Court of Justice for Ontario, *R. v. Menezes*,[46] Justice Hill summarized the case law as it related to the definition of "wanton" and "reckless",

> The term wanton means "heedlessly" (*R. v. Waite* (1996), 28 C.C.C. (3d) 326 (Ont. C.A.) at 341 per Cory J.A. (as he then was)) or "ungoverned" and "undisciplined" (as approved in *R. v. Sharp* (1984), 12 C.C.C. (3d) 428 (Ont. C.A.) at 430 per Morden J.A.) or an "unrestrained disregard for consequences" (*R. v. Pinske* (1988), 6 M.V.R. (2d) 19 (B.C.C.A.) at 33 per Craig J.A. (affirmed on a different basis [1989] 2 S.C.R. 979 at 979 per Lamer J. (as he then was)). The word "reckless" means "heedless of consequences, headlong, irresponsible": *R. v. Sharp, supra* at 30.[47]

Although these two judgments seem to share the same basic understanding of the meaning behind the use of the words "wanton" and "reckless", there have been differing views among the judiciary. For instance, Madame Justice Wilson in *R. v. Tutton*[48]found:

> The expression "wanton" disregard for the lives and safety of others is perhaps less clear. The word "wanton" taken in its acontextual sense could signal an ele-

---

45   *Ibid.* at 341.
46   (2002), 50 C.R. (5th) 343 (Ont. S.C.J.).
47   *Ibid.* at para. 72.
48   [1989] 1 S.C.R. 1436.

ment of randomness or arbitrariness more akin to an objective standard but, given the context in which it appears, coupled with the adjective reckless, and its clear use to accentuate and make more heinous the already serious matter of disregard for the lives or safety of others. I would think that the preferable interpretation is that the word wanton was intended to connote wilful blindness to the prohibited risk.

In short, the phrase "wanton or reckless disregard for the lives or safety of other persons" signifies more than gross negligence in the objective sense. It requires some degree of awareness or advertence to the threat to the lives or safety of others or alternatively a wilful blindness to that threat which is culpable in light of the gravity of the risk that is prohibited.[49]

Madame Justice Wilson's interpretation, although it seems to expand that of Justice Cory in *Waite*, was adopted by Justice Verville a year later in *R. v. I.W.F.*[50]

In the context of a corporation, the meaning of wanton or reckless disregard does not change from that of individuals. This matter was considered by Justice McKay in *R. v. Canadian Liquid Air Ltd.*[51] The meaning of the phrase was provided as follows:

"Reckless" means a person shows carelessness for the consequences of his act so far as the lives or safety of other persons are concerned. "Wanton" is a word with much the same meaning, but includes the idea not only of an indifference to consequences but of an unrestrained disregard of the consequences. A corporation can be reckless or wanton in its conduct or in a failure to act if its responsible officers for the action or omission in question have been reckless or wanton.[52]

Just as the fault element may be implied to the corporation by the employee's prohibited act, the wanton and reckless disregard was also, prior to Bill C-45, implied if the employee is found to be a directing mind of the corporation. Therefore, in considering the decision of Justice McKay in the context of post-Bill C-45 organizational guilt, it will have limited application.

CORPORATE CRIMINAL LIABILITY

Corporate criminal liability, prior to Bill C-45, was not a major focus of the *Criminal Code*. Developments in the area of the criminal accountability of corporations struggled to find an appropriate theory of how to hold a corporation criminally liable. A corporation is not a human

---

[49] *Ibid.* at 1406-7.
[50] (2000), 264 A.R. 319 (Alta.Q.B.).
[51] (1972), 20 C.R.N.S. 208 (B.C.S.C.).
[52] *Ibid.* at para. 6.

person, but rather a legal fiction established by statute. Due to a lack of clear direction provided by the *Criminal Code* on how to establish corporate criminal liability, the issue of how a corporation could be held liable for a crime was addressed in Canadian legal jurisprudence.

Corporations have been held to be criminally liable for acts that have been specifically authorized by the directing mind of the corporation. The corporation could not insulate itself from crimes committed by senior officers or directing minds of the corporation. This principle developed from the identification theory. Canadian courts have developed the identification theory of corporate criminal liability to hold a corporation responsible for the acts and omissions of senior officers or directing minds of the corporation. An early English case described this level of corporate decision maker as, "the directing mind and will of the corporation, the very ego and centre of the personality of the corporation".[53] The basis of corporate criminal liability in Canada goes back to the case of *R. v. Union Colliery Co. of British Columbia.*[54] The company had been indicted for unlawfully causing the death of certain persons by neglecting to properly maintain a bridge over which certain trains passed. The company was charged and convicted of unlawfully neglecting, without lawful excuse, to take reasonable precautions and to use reasonable care in maintaining the trust bridge in question. As a result, a locomotive engine and several cars broke through the bridge and fell into the valley of the Trent River, resulting in six deaths. The corporation was convicted at trial and eventually appealed to the Supreme Court of Canada. This 1900 Supreme Court of Canada case established that "a corporation can render itself amenable to the criminal law for acts resulting in damage to numbers of people, or which are invasions of the rights or privileges of the public at large, or detrimental to the general well being or interests of the state."[55] Notwithstanding this acknowledgement, the question remained as to the nature and scope of this liability, particularly with respect to the offence of criminal negligence.

Since the establishment of corporate criminal liability, the common law has evolved to somewhat broader concept of corporate responsibility.[56] Justice Schroeder in *St. Lawrence*[57] attributed this trend

---

[53] *Lennard's Carrying Co. Ltd. v. Asiatic Petroleum Co. Ltd.*, [1915] A.C. 705 at 713 (H.L.).

[54] (1900), 31 S.C.R. 81.

[55] *Ibid.* at 84.

[56] *R. v. St. Lawrence Corp. Ltd.* [1969] 3 C.C.C. 263 at para. 24 (Ont. C.A.).

[57] *Ibid.*

to the fact that, "corporations are at once more powerful and more materially endowed and equipped than are individuals and, if allowed to roam unchecked in the field of industry and commerce, they are potentially more dangerous and can inflict greater harm upon the public than can their weaker competitors."[58] In light of Bill C-45, it is clear that this principle will continue to expand, specifically in relation to the corporation's liability for *any* employee's actions that are covered by the new OHS legal duty added to the *Criminal Code*.

As a matter of some interest, the term corporation was not defined in the *Criminal Code*. Therefore, the definition of a corporation would be found in appropriate corporate law statutes under federal or provincial legislation. One of the important changes made by Bill C-45 is the replacement of the word "corporation" with the term "organization" in the *Criminal Code.* The detailed analysis of Bill C-45, in chapter 4 of this book, will review the definition of the new term organization and its implications for criminal law. Therefore, in the post-Bill C-45 era of the *Criminal Code,* it will be more accurate and appropriate to discuss organizational criminal liability rather than corporate criminal liability.

THE FAULT ELEMENT FOR ORGANIZATIONS AND CORPORATIONS

It is trite to say that modern Canadian law regulates organizations, including corporations, in a number of different ways. Corporations are regulated for purposes of registration, taxation, unfair business competition, price fixing, environmental compliance, public health concerns, and of course occupational health and safety. The latter subject of OHS regulation is covered in chapter 5 of this book.

Organizational criminal liability requires proof beyond a reasonable doubt of a prohibited or unlawful act and a fault element or *mens rea*. The fault element may be implied by the general requirements and criminal law jurisprudence or it may be more specifically found in the language of the *Criminal Code*. The offence of criminal negligence uses the fault element of wanton and reckless disregard.

Canadian courts have generally resisted holding organizations vicariously liable for the criminal offences of their employees. Rather, Canadian criminal law jurisprudence has developed the identification theory of organizational criminal liability. To establish the fault element

---

[58] *Ibid.*

required by a criminal law for an organization, a senior official or directing mind must be proven to have the fault element of the criminal offence and it will be thereby attributed to the organization because of the position of authority that individual had in its affairs. The terms organization and corporation are used interchangeably for the remainder of this section of this chapter. The corporation, however, will generally not be held liable if the senior official or directing mind was acting completely or fraudulently against the interests of the corporation. In that situation, the corporation itself would be a victim of the crime and not the criminal perpetrator. Canadian jurisprudence over time has to broaden the scope of the identification theory. However, it is not the only theory of corporate criminal liability. The identification theory and other models of corporate criminal liability are discussed later in this chapter.[59] The remainder of this section will address the issue of the pre-Bill C-45 fault element for corporate criminal liability in a charge of criminal negligence and some examples of its application to workplace health and safety criminal negligence prosecutions.

The fault element is a particularly important and problematic issue from the perspective of an organization. The general law relating to criminal negligence involving an individual is less concerned with this issue since the individual is without question responsible for their own actions.[60] However, it is much different when a criminal negligence charge involves a corporation. The question then becomes *who* in a corporation must have fault element so that the corporation itself can be said to have the fault element. This question is not difficult when the corporation is small and the owner is also the manager. In that case, the mind of the individual owner is also the mind of the corporation. However, large modern corporations may have organizational charts and structures that bear only a passing resemblance to the simpler models considered by the courts in developing the common law.

The Ontario Court of Appeal considered this issue in *R. v. Canadian Allis-Chalmers Ltd.*[61] In this case a large crane with a bucket designed to carry sand to another location malfunctioned and opened, pouring sand onto an unsuspecting employee causing serious injury and eventual death. The defendant corporation was a foundry of large dimensions employing about 175 workers. The foundry had three bays in each of

---

[59] See Chapter 3(G).
[60] As discussed in the previous section, the question of intent with regard to the individual is still unresolved.
[61] [1923] 54 O.L.R. 38 (Ont. C.A.)

which was a travelling overhead crane, which was used to life heavy loads and move them from one part of the bay to another. A large overhead crane, manufactured in the United States and used there, in Canada and elsewhere, was used to move a large bucket which moved sand from the sandpit in the foundry to the moulder. On one occasion, a worker was standing underneath the bucket, which was lifted by the crane, when the bucket dumped its load even though no one pulled the lever that would normally activate the dumping of the load. The worker suffered severe injuries, he then contracted pneumonia and eventually died.

The court delivered three concurring judgments dismissing the appeal from the acquittal at trial. Of particular interest is Justice Masten's conclusion that:

> In order that there may be a conviction there must, in my opinion, be no reasonable doubt that the company itself failed to take proper precautions against and use reasonable care to avoid the danger in question, and I think that the failure of the men operating the machine to report the apparent defect was not negligence of the company itself but personal negligence of these men. In negligently omitting to do so the workmen were acting in that respect contrary to the instructions of the company; and in omitting to report they were not acting for or on behalf of the company. Their negligence was not for the company's benefit, but in opposition to its instructions. The true position is that, through their own personal negligence and for their own convenience, they omitted to do what they ought to have done and what the company had ordered them to do.[62]

In Justice Orde's concurring judgment he suggested that a line be drawn between those in authority and those who are not in authority. Although he did not go on to actually draw the line in that case, but rather left it to the circumstances of each case, he stated that:

> I am not prepared to hold that the negligence of a minor servant of the company, even though he may be invested with some authority, such as that of a foreman over a gang of men, is to be regarded as the criminal negligence of the company. I can see no reason for placing a corporate body in any lower position in this regard than an individual.[63]

Although this case was decided in 1923, the principles have been consistently adopted in Canadian courts up to the passage of Bill C-45 into law.

Another important case of alleged criminal negligence involving a workplace accident is *R. v. Syncrude Canada Ltd.*[64] In that case, two

---

[62] *Ibid.* at para. 31.
[63] *Ibid.* at para. 41.
[64] (1983), 28 Alta.L.R. (2d) 233.

men, working for a contractor repairing a large tank, were asphyxiated due to exposure to nitrogen gas after going into a reactor in order to retrieve a fallen tool. The defendant, Syncrude Canada Ltd. had hired an independent contractor, Western Stress Relieving Servicing Inc. That contractor was hired to carry out repairs to the Syncrude plant near Mildred Lake, Alberta, during a plant shutdown. While servicing a reactor, in a confined space, a worker dropped a wrench. The worker descended into the reactor to retrieve the wrench and was overcome by toxic fumes. A co-worker, observing the fate of the first worker who was trying to retrieve the wrench, descended into the same reactor to help and was similarly overcome by the toxic fumes. The medical evidence at the trial indicated that both men were likely overcome and rendered unconscious in seconds by nitrogen gas in the reactor.

Syncrude was subsequently charged with criminal negligence causing death. Justice Agrios, the trial judge, after a thorough examination of the development of corporate criminal negligence, acquitted the accused corporation. The basis of that decision is informative and illustrates the pre-Bill C-45 law with respect to the fault element corporate criminal liability.

Justice Agrios concluded that it was necessary to look to the alter ego or directing mind of the corporation and that a "clear distinction is made between the acts of inferior agents or servants of the corporation, as opposed to the acts of more responsible officers of the company"[65]; the latter of which would fix the corporation with criminal liability. In that case the employees who were responsible for the prohibited act were processing technicians who issued safe work permits. Justice Agrios, in dismissing the charges of criminal negligence, stated that:

> I cannot find that in a company with 4,000 employees, permit issuers, albeit they have the authority to implement safety procedures, can be considered the directing mind and will of the corporation, the alter ego of the corporation…Their rank and position is not such of a nature as to justify the finding that their acts might be ascribed to the company itself so as to fix the corporation with liability for their acts or omissions.[66]

The fault element of criminal negligence derives from the individual who in fact committed the act or omission within the organization. Depending on whether or not the corporate representative falls within the directing mind test described above, their position and status determines the implied fault element of the offence against the corporation.

---

[65]   *Ibid.* at para. 40.
[66]   *Ibid.* at para. 48.

The standard of corporate liability for criminal negligence is demonstrated in the prosecution of *Canadian Liquid Air Limited*[67], a prosecution in British Columbia relating to a worker's death. The corporate accused made efforts to make the work area where the incident occurred free of any toxic gases. The corporation was aware that employees of a contractor would be using a cutting torch attached to an acetylene holding tank and that toxic gases must not be in the area when the torch was used. Unfortunately, the accused corporation did not succeed in making the area completely free of toxic gases, resulting in the death of a contract worker. Although the evidence established a clear breach of a duty, prosecution failed to prove beyond a reasonable doubt the fault element necessary to prove criminal negligence. Justice McKay stated that,

> ...it seems clear to me that a corporation can be in breach of either or both of the duties, and can breach the duty in a manner that shows a wanton or reckless disregard for the lives or safety of other persons...a corporation can be reckless or wanton in its conduct or in a failure to act if its responsible officers for the action or omission in question have been reckless or wanton.[68]

The procedure for purging the toxic gases from the area in which the contract workers were working was to introduce nitrogen gas into the area. An industrial hygienist who testified at the trial said that purging with nitrogen would be a suitable procedure to adopt to ensure that there was a safe work environment. However, it was clear from the end results and the death of the worker that the purging with nitrogen did not make the area gas-free. There appears to be some indication in this court decision that the worker who conducted the purging activity was negligent in carrying it out. However, the employer did not take tests after the purging, to determine the purging had been successful. The court held that the prosecution had not offered any evidence in the trial of inadequate purging techniques and no evidence of inadequate purging equipment. There was also no evidence of incompetent personnel involved in the purging and a lack of evidence on how the approved purging procedure failed. In the result, the court acquitted the corporate accused. This case would likely have been different and may have resulted in a conviction of the corporation if the new organizational liability provisions under the Bill C-45 amendments to the *Criminal Code* had been applied.

---

[67] (1972), 20 C.R.N.S. 208 (B.C.S.C.).
[68] *Ibid.* at 210.

MODELS OF CORPORATE CRIMINAL LIABILITY

The primary issue regarding the various models of corporate criminal liability is who in a corporation is responsible for implying liability onto the corporation. It is also critical to set out the process by which corporate criminal liability is established. Different jurisdictions have approached corporate liability through a variety of theories or models. Canada courts have, in the past, adopted the identification theory, whereas the vicarious liability model or *respondent superior* model is used in the United States. Australia has established the corporate culture model, and corporate killing legislation is currently being discussed in the United Kingdom. These models were considered in the public policy and political review process that ultimately culminated in Bill C-45. Each model will be briefly discussed below.

Identification Theory

The prevailing, pre-Bill C-45, model of corporate criminal liability in Canadian criminal jurisprudence is the identification theory. This theory holds corporations liable for the acts and omissions of senior officials and directing minds of the corporation. In the modern corporation there is some difficulty in establishing who is a senior official or a directing mind that will bind the corporation for the purposes of criminal liability. The directing mind normally must be held to have executive level authority in the corporation. However, depending on the size, structure and hierarchy of the corporation that is not always an easy level of authority to identify. The identification theory has come under criticism, for a number of years, for placing too many limits on who might be considered the corporation's directing mind. Since criminal law is restrictively applied, to safeguard the rights of the accused, a growing consensus determined that the identification theory was inadequate to address the modern, complex corporation.

The identification theory was recognized and clearly supported in Canadian law in *R. v. Canadian Dredge & Dock Co.*[69] This criminal prosecution related to alleged bid-rigging the dredging of the Hamilton harbour. Several charges in the indictments related to contracts between public authorities and the accused where the bids were alleged to have been tendered on a collusive basis, with the low bidders arranging to compensate high bidders or non-bidders in order to secure the contract.

---

[69] [1985] 1 S.C.R. 662.

Each company had a manager who conducted the business of the company relating to the submission for bids for tender. The corporate accused denied any criminal involvement because the managers were acting fraudulently against their own employers, were acting on their own behalf, and were acting contrary to instructions given to them. The identification theory focuses on the actions of the directing mind of the corporation and merges individual and corporate persons in order to assign criminal liability. The Supreme Court of Canada expanded the directing mind in Canadian Dredge & Dock Co. to include "the board of directors, the managing director, the superintendent, the manager or anyone else delegated by the board of directors to whom is delegated the governing executive authority of the corporation."[70] The target group for establishing corporate criminal liability under the identification theory in the common law was previously narrower, including only the board of directors, the managing director and other managers who are highly placed.[71]

A further example of the identification theory is found in the prosecution of *R. v. Waterloo Mercury Sales Ltd.*[72] In that case, an Alberta car dealer company was prosecuted for rolling back odometers and fraudulently representing to potential customers lower mileage of various motor vehicles. The manager of the used car lot was held to be the corporation's directing mind for the purposes of determining whether it had committed fraud. In this case, the president of the corporation was found to have had no knowledge of the conduct by the manager of the used car lot part of the business. The evidence at trial indicated that the president of the company had circulated written instructions prohibiting the rollback of odometers on used vehicles offered for sale. The court held that the manager was acting as the corporation's directing mind within the field of operation assigned to him, the sale of used cars. However, it appears that an employee with lesser responsibility, such as a used car salesperson or a mechanic committing the same fraudulent act of rolling back odometers, may not be held to be a directing mind of the corporation under the identification theory.

---

[70] *Ibid.* at para. 32.
[71] *Ibid.*
[72] (1974), 18 C.C.C. (2d) 248 (Alta. Dist. Ct.)

Vicarious Liability Model

Vicarious liability in tort law is the attribution to the corporation of a civil wrong by a representative of the corporation. Vicarious liability may occur when the acts or omissions of another person, such as a representative of the corporation, are held to be for the purposes of legal liability, the acts or omissions of the corporation. The general rule in criminal law was that courts had a presumption against vicarious liability unless there was clear legislative intention to establish criminal acts by vicarious liability. In Canadian criminal law, vicarious liability has three specific requirements. First, it must be established that the employee committed the crime with the requisite fault element, and if established it is imputed to the corporation. The fault element may also be established on the collective knowledge of the employees as a group. Second, the employee must have acted within the scope of his or her employment. Third, the employee must have intended to benefit the corporation in some way. The final two requirements have been criticized for being extremely broad and easily met, thereby imposing liability without significant restriction.[73]

The vicarious liability model has been considered and rejected in both Canada and the United Kingdom. However, the Government of Canada recognized that the differences between the identification theory and the vicarious model were not that significant. In its response to the Fifteenth Report of the Standing Committee on Justice and Human Rights, the federal government quoted Professor Healy to explain its position:

> The identification doctrine is vicarious liability. It's just at a very focused, narrow level, whereas you can have a wider range of people engaging the liability of the corporation under the American approach.[74]

Despite the appearance of a similarity between the vicarious liability model and the identification theory, Canada has been reluctant to go as far as the United States has in adapting the vicarious liability model, due to issues involving the *Charter*. Specifically, the concern was that an absolute liability offence would be created if the net was cast too

---

[73] Dean Jobb, *Calculated Risk—Greed, Politics, and The Westray Tragedy* (Halifax: Nimbus Publishing Ltd., 1994) at 3.

[74] Canada, Department of Justice, *Government Response to the Fifteenth Report of the Standing Committee on Justice and Human Rights* (Ottawa: Department of Justice, 2002) at 9, online: <http://canada.justice.gc.ca/en/dept/pub/ccl_rpm/>: For Professor Healy's original comments in context, see the *Minutes of Proceedings*, at 1205, on-line: http://www.parl.gc.ca/InfoComDoc/37/1/JUST/Meetings/Evidence/JUSTEV91-E.HTM>.

broadly, and that the *Charter* would not allow that result. This concern was addressed in the United Steelworkers of America's presentation to the Committee in their Bill C-45 submissions: "…things could be going on that people don't know anything about in the corporation, that, in fact, aren't even authorized by the corporation. In some jurisdictions in the U.S. that's still sufficient to warrant a conviction for the corporation. That … would run into a considerable number of *Charter* challenges, as well as not accomplishing the purpose."[75]

Corporate Culture Model

Bill C-284, discussed in chapter 2, proposed a model of corporate culture criminal liability. While the federal government did not ultimately follow this direction, Australia has adopted the corporate culture model of criminal liability. That model has four alternative elements:

(i)   the board of directors intentionally, knowingly or recklessly carried out the relevant conduct, or expressly, tacitly or impliedly authorized or permitted the commission of the offence; or

(ii)   a high managerial agent intentionally, knowingly or recklessly engaged in the relevant conduct, or expressly, tacitly or impliedly authorized or permitted the commission of the offence; or

(iii) a corporate culture existed within the corporation that directed, encouraged, tolerated or led to non-compliance with the relevant provision; or

(iv)   the corporation failed to create and maintain a corporate culture that required compliance with the relevant position.[76]

Canada's pre-Bill C-45 criminal law standard for corporations was represented by the first two elements of the model. The final two points represent a divergence from Canadian corporate criminal liability rules. By establishing corporate criminal liability Bill C-284 would have controlled management's activities if it:

---

[75]   From the testimony of Andrew King, Department Leader for Health, Safety and the Environment for the United States of America, online: <http://www.parl.gc.ca/ InfoComDoc/37/1/JUST/Meetings/Evidence/JUSTEV86-e/htm>.

[76]   Canada, Department of Justice, *Government Response to the Fifteenth Report of the Standing Committee on Justice and Human Rights* (Ottawa: Department of Justice, 2002) at 9, online: <http://canada.justice.gc.ca/en/dept/pub/ccl_rpm/>. See also the Australian *Criminal Code Act 1995* (Cth.) s. 12.1(2).

(i)   tolerated, condoned or encouraged the act by the policies or practices it permitted,

(ii)   was wilfully blind to the act or omission,

(iii)   had allowed the development of a culture among its officers and employees that encouraged them to believe that the act or omission would be tolerated, or

(iv)   failed to take steps that a reasonable corporation should take to ensure its employees knew the act or omission was unlawful or forbidden by the corporation.[77]

While some witnesses at the Committee hearings were in favour of a corporate culture model, the Canadian government decided that the model would not necessarily achieve its intended results. Furthermore, the Australian model had not yet been applied in any criminal cases. Therefore in the Canadian government's view the law remained an untested basis for criminal liability. The government response to the Committee report said:

> The Government is conscious of the need for clarity in the law and considers that "corporate culture" is too vague to constitute the necessary corporate *mens rea*. Any changes to the criminal law should be simpler and depart less from general principles than the corporate culture model.[78]

One of the most significant detractions of this model seems to be that it would be possible for a corporation to be held criminally liable without there being a requirement for an individual within the corporation to be found guilty of an offence. The federal government was not prepared to adopt this broad collective responsibility approach to corporate criminal liability.

Corporate Killing Model

The corporate killing model is an example of a specific criminal offence that focuses on a specific type of prohibited activity, workplace injury and death. Arguably this is an appropriate description of Recommendation 73 of the Westray Inquiry. This model only addressed corporate criminal liability in the workplace health and safety field. It was considered too narrow in Canada, as a general theory of corporate criminal liability, as it

---

[77]   *Ibid.*

[78]   *Ibid.* at 11.

would not respond to corporate crimes against the environment and other corporate harm.

There are, at the time of writing, legislative proposals in the United Kingdom and Australia dealing with corporate manslaughter and worker health and safety. The Committee heard witnesses in support of and against specific corporate killing legislation. On the one hand, it would remove the corporate veil that protects individuals; however, on the other hand, it forms the possibility of "creating anecdotal legislation that does not cover the entire terrain."[79] One witness is cited in the Government Response as being in favour of the specific corporate killing legislation because in writing such a law "rather than rewriting the law of corporate criminal liability...[it] probably invites less legal and constitutional peril."[80]

The federal government chose to use Bill C-45 to rewrite the law of corporate criminal liability and intentionally did not have a corporate killing offence in Canada. Even though this model was rejected by Canadian law makers, the British initiative in this direction is still alive. According to a July 9, 2003 online BBC report, David Bergman of the Centre for Corporate Accountability says the British government is still planning to enact a new offence of corporate killing:

> This would allow any employing organization to be prosecuted for causing a death as a result of a very serious management failure on the part of the organization. It will make it easier to prosecute a company for a homicide offence.[81]

The British government has not, however, committed to a specific timetable when the bill will be introduced before parliament.

---

[79] *Ibid*. at 9.
[80] *Ibid*.
[81] BBC News, "Q&A: Corporate manslaughter Network Rail and Balfour Beatty have been charged with corporate manslaughter over the Hatfield rail crash of October 2000" (July 9, 2003) , online: <http://news.bbc.co.uk/1/hi/uk/3053239.stm>.

CHAPTER 4

DETAILED ANALYSIS OF BILL C-45

A full understanding of Bill C-45, including its interpretation and implications, requires a detailed analysis of the legislation. In each of the sections that follows, the applicable provision of Bill C-45 is set out, then the provision is analyzed. Each of the section headings deals with a separate section of Bill C-45. In each specific section, either an existing provision of the *Criminal Code* is amended, altered, or replaced or, alternatively, a new provision, such as section 217.1, is added for the first time to the *Criminal Code*. As can be seen from the detailed review of the following sections, Bill C-45 changes a number of provisions of the *Criminal Code*. The following analysis is intended to identify, section by section, the changes to the *Criminal Code* under Bill C-45 and their importance to other provisions under Bill C-45, and the *Criminal Code* generally. The detailed analysis of Bill C-45 in each section and taken together provide a clearer picture of the purposes and effects of the new legislation. At the time of writing none of these provisions have been judicially considered. In a future edition of this book, a review of jurisprudence with respect to these provisions will be provided.

SECTION 1: EXTENDS POTENTIAL CRIMINAL LIABILITY

1.(1) – The definition " "every one", "person", "owner" " in section 2 of the *Criminal Code* is replaced by the following:

"every one", "person" and "owner", and similar expressions, include Her Majesty and an organization;

1.(2) – Section 2 of the Act is amended by adding the following in alphabetical order:

"organization" means

(*a*)   a public body, body corporate, society, company, firm, partnership, trade union or municipality, or

(*b*)   an association of persons that
   (i)   is created for a common purpose,

   (ii)   has an operational structure, and

   (iii)   holds itself out to the public as an association of persons;

"representative", in respect of an organization, means a director, partner, employee, member, agent or contractor of the organization.

"senior officer" means a representative who plays an important role in the establishment of the organization's policies or is responsible for managing an important aspect of the organization's activities and, in the case of a body corporate, includes a director, its chief executive officer and its chief financial officer.

Section 1 of Bill C-45 substantially amends a number of terms that are defined under the *Criminal Code*. The terms "organization", "representative", and "senior officer" were not previously used or defined in the *Criminal Code*. Prior to Bill C-45, the *Criminal Code* stated:

"Every one", "person", "owner", and similar expressions include Her Majesty and public bodies, bodies corporate, societies, companies and inhabitants of counties, parishes, municipalities or other districts in relation to the acts and things that they are capable of doing and owning respectively.[1]

Section 1 transfers much of the definition of the terms "every one", "person", "owner" and similar expressions to the new definition of organization. By establishing the new term "organization", and incorporating it by reference into the definition of every one, person, owner, and similar expressions, these definitions have been simplified. The focus now under Bill C-45 is clearly on the definition and application of the new term "organization".

Bill C-45 broadened the common law approach to corporate criminal liability by expressly extending criminal negligence beyond corporations and companies to any organization. The rationale behind this change is to ensure that the same rules for attributing criminal liability apply to various forms of organizations, regardless of how they choose to organize themselves or structure their affairs. Now public bodies, trade unions, municipalities and associations of persons have the same potential criminal liability as corporations. The effect will be to cast the net of potential criminal liability wider, extending it to organizations

---

[1]   R.S.C. 1985, c. C-46.

whether they are incorporated or not, operate for profit or not, are publicly listed or not.

Bill C-45 changes the criminal law with respect to the liability of organizations. It was intended to lower the threshold by which a prosecutor must prove the criminal offence against an organization. The specific types of bodies that are included in this new definition of organization are as follows: public bodies, corporate bodies, societies, companies, firms, partnerships, trade unions and municipalities or any association of persons that is created for a common purpose, that has an operational structure and that holds itself out to the public as an association of persons.

There is no statutory exemption for non-profit organizations, charitable organizations, or religious organizations. The debates in the House of Commons and the Senate completely ignored this implication of Bill C-45. Any organization, as defined in Bill C-45, may now be held criminally liable for its actions or omissions if they constitute an offence under the *Criminal Code.*

As will be analyzed later in this chapter, the new terms of representative and senior officer have specific importance in the new formulas for organizational criminal liability. A representative of an organization includes a director, partner, employee, member, agent or contractor. A senior officer includes anyone who plays an important role in the policy making of the organization or anyone who is responsible for managing important activities. In the case of a corporation, it includes a director, Chief Executive Officer or Chief Financial Officer. These two terms are instrumental in the new statutory formulas for the fault element to establish organizational guilt reviewed.

SECTION 2: EXTENDS POTENTIAL CRIMINAL LIABILITY DEEPER INTO THE ORGANIZATIONAL STRUCTURE

2. The Act is amended by adding the following after section 22:

22.1 In respect of an offence that requires the prosecution to prove negligence, an organization is a party to the offence if

    (*a*) acting within the scope of their authority

        (i)   one of its representatives is a party to the offence, or

        (ii) two or more of its representatives engage in conduct, whether by act or omission, such that, if it had been the conduct of only one representative, that representative would have been a party to the offence; and

(b) the senior officer who is responsible for the aspect of the organization's activities that is relevant to the offence departs – or the senior officers, collectively, depart – markedly from the standard of care that, in the circumstances, could reasonably be expected to prevent a representative of the organization from being a party to the offence.

Section 2 introduces two new formulas by which organizations may be held criminally liable. These two formulas are a departure from the identification theory of corporate liability. The first, section 22.1 of the *Criminal Code*, is in respect of offences that required proof of negligence. This section applies to the new OHS legal duty in section 217.1, that will be reviewed later in this chapter, as well as other forms of criminal negligence. This provision also recognizes that not all criminal offences have the same fault element. Section 22.2 deals with the new formula for establishing organizational guilt where the fault element is not negligence. In a very real sense, section 22.1 established a statutory fault element that is now the measure of negligence to establish the new offence of OHS criminal negligence, or any other criminal negligence charge. There is no requirement to rely on theories of liability developed by the courts. However, courts will still interpret and apply this statutory fault provision. In other words, new section 22.1 provides a formula for proving organization guilt for an offence with a fault element of negligence, even if it is unrelated to the new OHS legal duty in section 217.1 of the *Criminal Code*.

This section sets out when an organization may be held liable for criminal negligence for the acts of its representatives and senior officers. Under the law prior to Bill C-45, a corporation could be found liable for a *Criminal Code* offence, but it was left to the common law to determine the nature and breadth of this liability. In the past, individuals within corporations have been found criminally liable by a directing mind, such as officers and directors, committing the criminal offence. Under the identification theory the guilt or innocence of a few senior corporate executives and directors is what determined the guilt or innocence of the corporation. However, the new definitions of representative and senior officer list the individuals in this role and how they may contribute to the finding of criminal liability by the organization.

Under the new *Criminal Code* amendments, an organization will be considered a party to the offence if one or more representatives, acting within the scope of their authority, is a party to the offence; and, if a senior officer responsible for the activity departs markedly from the standard of care expected to prevent that representative from being a party to the offence. This standard of criminal liability is extended to all organizations, not just corporations.

The Supreme Court of Canada in *R. v. Creighton*[2] set the standard with respect to determining criminal liability of corporations. In that case, Madam Justice McLachlin stated that the test for whether the accused's conduct constituted a marked departure from the standard of care of a reasonable prudent person in the same circumstances should not be extended to incorporate a standard of care which varies with the background and predisposition of the accused. That objective legal test has largely been adopted in Bill C-45 by the inclusion of section 22.1 into the *Criminal Code*.

> **22.2** In respect of an offence that requires the prosecution to prove fault – other than negligence – an organization is a party to the offence if, with the intent at least in part to benefit the organization, one of its senior officers
>
>   (*a*)   acting within the scope of their authority, is a party to the offence;
>
>   (*b*)   having the mental state required to be a party to the offence and acting within the scope of their authority, directs the work of other representatives of the organization so that they do the act or make the omission specified in the offence; or
>
>   (*c*)   knowing that a representative of the organization is or is about to be a party to the offence, does not take all reasonable measures to stop them from being a party to the offence.

The second part of section 2 of Bill C-45 deals with the formula for criminal liability where the fault element is a more traditional *mens rea* element associated with crimes. The fault element may be implied by the language of the criminal offence, as interpreted by jurisprudence, or, alternatively, set out quite clearly in the language of the offence. The difficulty in establishing criminal liability of corporations in a more traditional criminal charge was intended to be addressed by the addition of section 22.2 to the *Criminal Code*. This new provision expressly requires that the organization intended to benefit, at least in part, from the commission of the offence, to prove the criminal offence. Further, the new definition of senior officer in section 1 of Bill C-45 is critical in this new fault element to the *Criminal Code* as well as section 22.1.

This section sets out a new formula to establish when an organization may be held liable for criminal offences other than those with a fault element of negligence. An organization will be considered a party to an offence if a senior officer, with the intent, at least in part to benefit the

---

[2]   [1993] 3 S.C.R. 3.

organization and acting within the scope of his or her duty, either is (a) a party to the offence; (b) directs work to other representatives so that they commit the offence; or, (c) knowing that a representative is or is about to be a party to the offence, fails to take all reasonable measures to stop them. The fault element is disjunctive, establishing three different means by which the organization may be found guilty of a criminal offence. If any one of these three alternative scenarios occurs, organizational criminal liability will be established.

SECTION 3: ESTABLISHES A HEALTH AND SAFETY DUTY FOR PERSONS DIRECTING WORK OR AUTHORIZED TO DO SO.

3. The Act is amended by adding the following after section 217:

217.1 Every one who undertakes, or has the authority, to direct how another person does work or performs a task is under a legal duty to take reasonable steps to prevent bodily harm to that person, or any other person, arising from that work or task.

Bill C-45 introduces, for the first time in Canadian legal history, a duty relating to occupational health and safety in the *Criminal Code*. The OHS legal duty applies to "every one ", which includes Her Majesty and an organization. An individual, an organization, and the federal and provincial governments have this legal duty. Further, the duty applies to every one who undertakes to direct how another person does work or performs a task. The use of the word "how", may be a restrictive modifier in determining the scope of this duty. In other words, it is not enough that a person undertakes, or has the authority, to direct another person to do work. This legal duty only applies if the person directs how the other person does the work or performs the task. The use of the word "how", an adjective, clearly modifies and restricts a set of individuals to whom the new legal duty applies. "How" is synonymous with the phrase "by what means", or "in what manner or way". The canon of statutory interpretation instruct courts that every word in a statute is intended to have meaning and should be given its plain and ordinary meaning. The word "how", therefore, modifies the phrase "another person does work or performs a task". Section 217.1 therefore establishes an OHS legal duty for every one who directs how, or the manner in which, the work is done.

Further, this legal duty also applies to every one who has the authority to direct how another person does work or performs a task. The application of this duty to every one who has the authority to direct how another person does work or performs a task extends it beyond the individual who directed the work or a task. The phrase, "or has the authority", is problematic because it does not provide a particular title or level of responsibility within the organization. It appears that any person who has the authority to direct how another person does work or performs a task is a much broader description of organizational decision-makers than the identification theory. Further, since the new Bill C-45 term "senior officer" is not used in section 217.1, it must be reasonably inferred that the legislature intended a different and broader set of individuals than those specifically defined in the term "senior officer" to have this new legal duty. What is clear from the phrase "or has the authority", in section 217.1, is that a court would have to specifically and carefully review who has the authority to direct how another person does work or performs tasks in order to establish the application of the new legal duty. Therefore, although this phrase gives some uncertainty as to how high in the organizational hierarchy a person may be who has the authority to direct how another person does work or performs tasks, it nevertheless gives trial courts a degree of flexibility to look at the specific organizational structure, and reporting authority, of each organization that is brought before the courts.

Express reference to the existence of the new "legal duty", in section 217.1, clearly establishes a basis for breach of this duty to amount to criminal negligence. Criminal negligence, as discussed in chapter 3, results from either an act or an omission relating to a legal duty, when a person shows wanton or reckless disregard for the lives or safety of other persons. Section 217.1 establishes a new legal duty, the breach of which may give rise to a charge of OHS criminal negligence.

The new legal duty is to take "reasonable steps" to prevent bodily harm to that person, the person to whom direction is given as to how to do the work or perform the task. The term "reasonable steps" is not defined. It is the writer's view that reasonable steps, at minimum, would refer to compliance with applicable OHS statutes and regulations. Reasonable steps may also refer to industry standards, codes of practice and in some cases best practices. Some further discussion of "reasonable steps" is undertaken in chapter 5 of this book.

The phrase "that person" is not specifically defined. This phrase, presumptively, would be an employee or a worker. The phrase "that person" in section 217.1 is broader than either the word "employee" or "worker", but would apply to both. There is no requirement for an employment relationship for this duty to apply. However, that phrase may extend to a volunteer or visitor who is the subject of a direction on law to perform work or a task.

Further, the legal duty extends to "any other person" in section 217.1. That phrase presumably would be as broad, as the summary of Bill C-45 suggests, to include the public. Therefore, the Bill C-45 new OHS legal duty extends to the public as well as workers. The phrase "any other person" is modified by the last subordinate clause in section 217.1 and states "arising from that work or task". Therefore, the legal duty to take reasonable steps to prevent bodily harm arises from and in relation to that work or task. It does not arise, in other words, unless the work or task or activity related to it causes the risk of bodily harm. In the writer's opinion, there must be evidence of a nexus or causal connection between the assignment of how "another person" does the work or performs a task and the risk of bodily harm to "that person" or "any other person" arising from that work or task.

For example, if several adolescents trespass onto a construction site, on an evening or a weekend when no construction activity was taking place, and they exposed themselves to hazards which resulted in bodily injury, those circumstances are not likely be covered by the section 217.1 duty. In other words, the legal duty to take reasonable steps is in relation to those persons that have been directed how to do the work or task and the other persons directly affected by the work or task being performed. In the writer's opinion, section 217.1 was not meant to apply to a trespasser who, without legal authority, puts themselves in proximity to a hazard in the workplace that does not directly arise from how the work or task is performed.

This new legal duty appears to borrow language from both OHS statutes and also the jurisprudence relating to the defence of strict liability offences known as due diligence. In chapter 5 of this book, there is a review of Canadian OHS law and general duty clauses in particular. OHS general duty clauses are similar in their language, purpose and intent to section 217.1 of the *Criminal Code*. Further, the language of the defence of due diligence, available for strict liability offences, makes references to the phrase "reasonable steps". The origin, nature, and legal test of due diligence is also set out in chapter 5.

In contrast to OHS statutes, regulations and the defence of due diligence, section 217.1 does not provide any specific steps, actions, control requirements or OHS systems that did not exist prior to Bill C-45 becoming law. In other words, although a new legal duty is added to the *Criminal Code* under section 217.1, it does not necessarily add any new specific OHS steps, actions, control requirements or systems for workplaces in Canada. In fact, the new legal duty to take reasonable steps to prevent bodily harm is best understood as reinforcing existing OHS statutes and regulations rather than amending, altering, or replacing them. Although applicable Canadian OHS statutes and regulations have their own regulatory enforcement mechanism, the rationale for adding this legal duty to the *Criminal Code* is to enhance the importance of accident and injury prevention, reinforce the legal seriousness by introducing a legal duty, and increasing the penalties and deterrents for failure to take reasonable steps to prevent workplace bodily harm under the *Criminal Code*. This was made clear in the Committee recommendations and the legislative debates regarding Bill C-45.

This legal duty gives rise to the potential of OHS criminal negligence charges and a number of related criminal offences. Under sections 21 to 24 of the *Criminal Code*, persons may be found criminally liable as parties to an offence by committing the offence and aiding, abetting, counselling, attempting, or being an accessory to the offence. Under these provisions, a corporate executive or board member may be liable if he or she aided or abetted a person to commit an offence in section 21, counselled a person to be a party to an offence in section 22, or was an accessory after the fact to an offence in section 23. It is important to understand that for each of these types of liability, the person is liable on account of his or her own actions.

Bill C-45 has increased the potential criminal liability of a greater number of individuals by creating a new legal duty and casting the net of potential liability deeper into the organizational structure and wider to include more individuals and all organizations. This amendment establishes a very broad and overarching legal duty to protect workers and the public at large. The legal duty, if not complied with, will amount to the prohibited act of the offence of criminal negligence.

Individuals who direct how another person does work within the organization must now realize that there is a criminal penalty for failure to properly discharge the legal duty. The legal duty extends from the most senior organizational decision-makers all the way to foremen, lead hands, and potentially even co-workers who direct how work or tasks are performed. This means that workers who would not normally expect to owe a duty to prevent harm in the workplace may now find themselves guilty of criminal negligence if they do not fulfill this duty and bodily harm or death results. All persons who undertake to direct how another person does work or performs a task, or has authority to do so, must be prepared to recognize and reasonably deal with health and safety hazards in the workplace in the face of this new legislation. The importance of OHS prevention has taken on a whole new importance and meaning.

However, the full implications of this new *Criminal Code* legal duty are, in some respects, unclear. The Bill C-45 amendments fail to define at least three key components of section 217.1. They are as follows: (1) "has the authority", (2) "reasonable steps", and (3) "any other person". This lack of guidance may lead to confusion as to the extent of the legal duty owed by every one, especially organizations. The subordinate clause "has the authority", from section 217.1 of the Bill C-45 amendments broadens the duty beyond just those persons who undertake to direct how another person does work or performs the task. Whether or not a person has actually directed how another person does work or performs a task, if they had the authority to so direct how another person does work or performs the task then the legal duty applies to them. In the writer's view it is unclear whether, for example, a non-employee volunteer director of a charitable organization would actually have the authority to direct how an employee of that non-profit corporation does work or performs tasks. This problematic subordinate clause will require a factual examination and determination of the positional and practical authority of that director.

The phrase "reasonable steps" is also not defined. Organizations will be looking for concrete statements of what constitutes "reasonable steps" to protect their workers and the public to avoid charges of OHS criminal negligence. The simple question is how can steps be determined to be reasonable? In the writer's view, it is likely sufficient to suggest that OHS statutory and regulatory compliance is sufficient when the consequences of missing the mark is so severe.

The phrase "any other person" is intended to extend the duty broader than just to employees or workers and likely intended to extend to public at large. If any member of the public is subject to bodily harm as a result of a work or task then the requirement to take reasonable steps applies. The members of the public that would most likely be affected would be visitors, students at take-a-student-to-work-day situations, and those members of the public that are affected by work on construction sites and other workplaces that have close proximity to public thoroughfares and transit systems. However, it remains unclear how closely connected or affected the public must be for the duty to apply. The phrase is so broad that it may be subject to attempts in public disasters to unreasonably broad or excessive application.

While it is clear that the intent of the legislation is to protect workers from harm in the workplace, it is also unclear what level of harm will trigger an offence under the *Criminal Code*. OHS legislation usually defines certain types of injury for purposes of reporting to the regulator. However, contravention of OHS legislation is possible without an accident or injury. This further complicates the task of determining what "reasonable steps" must be taken by the organization. The *Criminal Code* provides the following definition of bodily harm:

> ...any hurt or injury to a person that interferes with the health or comfort of the person and that is more than merely transient or trifling in nature.[3]

This very broad definition of bodily harm may give rise to a criminal prosecution for serious injuries and fatalities at the workplace. The jurisprudence relating to bodily harm was reviewed in chapter 3 of this book. This broad definition may also give rise to potential complaint and criminal prosecution for occupational disease, heat stress, and possibly even repetitive strain injury. It may be difficult to determine if bodily harm is more than merely transient or trifling in nature. However, in the writer's opinion, if a workers' compensation board recognizes either an injury or disease as a compensable claim, then it would likely amount to bodily harm as defined in the *Criminal Code*.

Finally, an employer will have to give considerable foresight into what a reasonable worker or the public would expect to encounter in respect of their workplace. The extent of this new duty is not certain in the sense that the work area may be changing or not easily defined. This is particularly important in construction projects with easy access to the public. It is also difficult to determine the full scope of all of the persons who will potentially have access to or be affected by the work or task. For example, will an employer be held criminally liable even after work has been completed and a member of the public is harmed by a condition resulting from the work or task that was completed weeks or months earlier? Remoteness of injury or bodily harm arising from that work or task is another issue that is not adequately addressed in this section. These questions will be resolved in the courts.

SECTION 4: REPLACES "CORPORATION" WITH "ORGANIZATION" WITH RESPECT TO THE OFFENCE OF THEFT.

4. Paragraph 328(e) of the Act is replaced by the following:

(*e*)   by the representatives of an organization from the organization.

---

[3]   R.S.C. 1985, c. C-46, s. 2.

Section 328 of the *Criminal Code* deals with theft by or from a person having a special property or interest in the thing stolen. This section describes the circumstances in which the offence of theft may be committed by the owner of an item or a person who has a special property or interest in the item that is stolen. For example, an equitable interest in the shares of a company resulted in a prosecution under this section, alleging that the accused had converted the shares of a company for their own personal use.[4] The Bill C-45 amendments of section 328 only changes one of the five special circumstances that apply to this crime. The amendment replaces "corporation" with the new term "organization" in paragraph 328(e) to broaden the situations in which this offence may arise. Prior to the Bill C-45 amendments, subparagraph 328(e) stated, "by the directors, officers or members of a company, body corporate, unincorporated body or of a society associated together for a lawful purpose from the company, body corporate, unincorporated body or society, as the case may be".

SECTION 5: REPLACES "CORPORATION" WITH "ORGANIZATION" WITH RESPECT TO FALSE PRETENCES.

5.(l) The portion of paragraph 362(1)(c) of the Act before subparagraph (i) is replaced by the following:

 (*c*) knowingly makes or causes to be made, directly or indirectly, a false statement in writing with intent that it should be relied on, with respect to the financial condition or means or ability to pay of himself or herself or any person or organization that he or she is interested in or that he or she acts for, for the purpose of procuring, in any form whatever, whether for his or her benefit or the benefit of that person or organization.

5.(2) Paragraph 362(1)(d) of the Act is replaced by the following:

 (*d*) knowing that a false statement in writing has been made with respect to the financial condition or means or ability to pay of himself or herself or another person or organization that he or she is interested in or that he or she acts for, procures on the faith of that statement, whether for his or her benefit or for the benefit of that person or organization, anything mentioned in subparagraphs (*c*)(i) to (vi).

Section 362 of the *Criminal Code* establishes the offence of making false statements or false pretences that are relied on for the purpose of financial decisions. This offence would occur when false information is

---

[4] *R. v. Ben Smith*, [1963] 1 C.C.C. 68 (Ont. C.A.)

given to a financial institution to secure a loan or mortgage. This section describes the elements of offence of false pretences and the appropriate punishment for such an offence. Subsection 362(1) identifies various means by which the offence of false pretence or false statement may be committed. This *Criminal Code* offence again is amended by the use of the new term "organization", as defined and discussed earlier in this chapter. This Bill C-45 amendment replaces the use of "corporation" with the new term "organization" as well as providing for both male and female persons in paragraphs 362(1)(c)and (d).

SECTIONS 6 AND 6.1: REPEALS SECTION 391 AND AMENDS SUBSECTION 418(2) OF THE *CRIMINAL CODE*

6. Section 391 of the Act is repealed

Section 391 of the *Criminal Code* provided that when a person acting in the name of a corporation, firm or partnership commits the offences of using misleading receipts, section 388, fraudulent disposal of goods on which money is advanced, section 389, or making fraudulent receipts under the *Bank Act*,[5] section 390, no person other than the person who is privy to the offence, is guilty of the offence. In other words, this section previously made it clear that in proceedings under sections 388, 389 or 390, only the person who actually committed an offence, or who is secretly privy to the commission of the offence, can be held accountable notwithstanding that a person is acting in the name of a corporation, firm or partnership. By repealing this section of the *Criminal Code*, Bill C-45 opens the door for organizations being held criminally responsible for such acts. The net effect of repealing section 391 is that both individuals and organizations may now be responsible for the criminal offences found in section 388, 389, and 390 of the *Criminal Code*. This is a substantive rather than procedural amendment to the *Criminal Code*.

6.1. The portion of subsection 418(2) of the Act before paragraph (a) is replaced by the following:

(2) Every one who, being a representative of an organization that commits, by fraud, an offence under subsection (1),

Section 418 of the *Criminal Code* establishes an offence of selling defective stores to Her Majesty or commits fraud in connection with the sale, lease or delivery of stores to Her Majesty or the manufacturer of

---

[5]   S.C. 1991, c. 46.

stores for Her Majesty. The *Criminal Code* establishes an indictable offence when this occurs. The term "Her Majesty" is defined in the *Interpretation Act.*[6] This Bill C-45 amendment is procedural and again relates to the new term introduced into the *Criminal Code* organization. This amendment replaces the use of "corporation" with the new term "organization", as well as replaces "director, an officer, an agent, or an employee" with "representative".

SECTION 7: REPLACES "CORPORATION" WITH "ORGANIZATION" WITH RESPECT TO PERSONS DEEMED TO HAVE ABSCONDED

7. Paragraph 462.38(3)(*b*) of the Act is replaced by the following:

> (*b*)   a warrant for the arrest of the person or a summons in respect of an organization has been issued in relation to that information, and

Section 462.38 of the *Criminal Code* provides for a forfeiture hearing for proceeds of crime in certain delineated circumstances. The appropriate attorney general for the jurisdiction that wants to take this action may apply for such a hearing if an information has been laid in respect of the designated criminal offence. Upon such application, the court may order forfeiture if the attorney general establishes, beyond a reasonable doubt, that the property in question is a proceed of crime, that related crime proceedings were commenced, and that the accused has either died or absconded from the jurisdiction. Subsection 462.38(3) is a provision that deems a person to have absconded if an information has been laid, a warrant issued, and reasonable attempts to arrest the person have been unsuccessful during a six-month period after the issuance of the warrant. The Bill C-45 amendment is in respect of the new term added to the *Criminal Code*, "organization" replacing the old term "corporation". With regard to warrants, a summons or warrant may be served on the organization rather than merely a corporation.

SECTION 8: REPLACES "CORPORATION" WITH "ORGANIZATION" WITH RESPECT TO PROCEDURE ON PRELIMINARY INQUIRY.

8. Section 538 of the Act is replaced by the following:

---

[6]   R.S.C. 1985, c. I-21, s. 35.

538. Where an accused is an organization, subsections 556(1) and (2) apply with such modifications as the circumstances require.

Section 538 of the *Criminal Code* permits a court to proceed with a preliminary inquiry in the absence of the accused organization. Prior to Bill C-45, this provision stated that a corporation that was duly served with a summons and did not appear, either by counsel or agent, may have its preliminary inquiry proceed in the absence of the accused corporation. Bill C-45 does not change the right of a corporation to appear by legal counsel or agent. It does, however, broaden the application of this principle to organizations, which includes corporations. The new term organization replaces corporation in this section with respect to appearances and the effect of non-appearance on a preliminary inquiry.

SECTION 9: REPLACES "CORPORATIONS" WITH "ORGANIZATION" WITH RESPECT TO APPEARANCES AND ELECTIONS.

9. Section 556 of the Act is replaced by the following:

556. (1) An accused organization shall appear by counsel or agent.

(2) Where an accused organization does not appear pursuant to a summons and service of the summons on the organization is proved, the provincial court judge or, in Nunavut, the judge of the Nunavut Court of Justice

   (*a*)   may, if the charge is one over which the judge has absolute jurisdiction, proceed with the trial of the charge in the absence of the accused organization; and

   (*b*)   shall, if the charge is not one over which the judge has absolute jurisdiction, hold a preliminary inquiry in accordance with Part XVIII in the absence of the accused organization.

(3) If an accused organization appears, but does not elect when put to an election under subsection 536(2) or 536.1(2), the provincial court judge or judge of the Nunavut Court of Justice shall hold a preliminary inquiry in accordance with Part XVIII.

Section 556 of the *Criminal Code* authorizes a court to take action in setting down an accused for trial or preliminary inquiry when the accused fails to appear by legal counsel or agent. Prior to Bill C-45, an accused corporation charged with a criminal offence must appear either by counsel or agent. The *Criminal Code* provided that where an accused corporation did not appear pursuant to a summons, and service of the summons is proven in court, a provincial court judge may proceed to trial in the absence of the accused if the criminal charge is within his absolute discretion. Further, where the criminal charge is not one over which the provincial court judge has absolute jurisdiction, the provincial court judge is required to fix a date for the trial or a date upon which the corporation must appear in trial court to have the trial date fixed. If, however, the corporation appears but has not requested a preliminary inquiry, the justice or judge, as the case may be, will be able to set a date for trial or a date for when the corporation is to appear in the trial court to fix a date. Bill C-45 amends this provision of the *Criminal Code* by adding the new term "organization" and replacing "corporations". Section 556 is amended to make those provisions regarding appearances and elections apply to organizations rather than just corporations. If an accused fails to appear and service of a summons is proved, the judge may either proceed with the trial, or hold a preliminary inquiry in the absence of the organization.

SECTION 10: REPLACES "CORPORATION" WITH "ORGANIZATION" WITH RESPECT TO WARRANTS OF COMMITTAL.

10. Subsection 570(5) of the Act is replaced by the following:

(5) Where an accused other than an organization is convicted, the judge or provincial court judge, as the case may be, shall issue or cause to be issued a warrant of committal in Form 21, and section 528 applies in respect of a warrant of committal issued under this subsection.

Section 570 of the *Criminal Code* sets out certain procedural steps to be taken when an accused is convicted. The *Criminal Code* provides for a record of conviction after a finding of guilt. Ordinarily where an accused is convicted the judge or provincial court judge shall also issue a warrant of committal using Form 21 under the *Criminal Code*. However, corporations pursuant to section 570(5) are exempt from such warrants of committal. Bill C-45 amends this provision to recognize the new term "organizations" in the *Criminal Code*. With respect to warrants of committal, this section remains unchanged with the exception that the new term "organization" replaces the term "corporations" for the exemption from the warrant of committal. An organization is exempt from the provisions of this section since an organization, including a corporation, cannot be committed for incarceration.

SECTION 11: REPLACES "CORPORATION" WITH "ORGANIZATION" WITH RESPECT TO APPEARANCES AND NOTICE

11. The heading before section 620 and sections 620 to 623 of the Act are replaced by the following:

Organizations

620. Every organization against which an indictment is filed shall appear and plead by counsel or agent.

621. (1) The clerk of the court or the prosecutor may, where an indictment is filed against an organization, cause a notice of the indictment to be served on the organization .

(2) A notice of an indictment referred to in subsection (1) shall set out the nature and purport of the indictment and advise that, unless the organization appears on the date set out in the notice or the date fixed under subsection 548(2.1), and enters a plea, a plea of not guilty will be entered for the accused by the court, and that the trial of the indictment will be proceeded with as though the organization had appeared and pleaded.

622. Where an organization does not appear in accordance with the notice referred to in section 621, the presiding judge may, on proof of service of the notice, order the clerk of the court to enter a plea of not guilty on behalf of the organization, and the plea has the same force and effect as if the organization had appeared by its counsel or agent and pleaded that plea.

623. Where an organization appears and pleads to an indictment or a plea of not guilty is entered by order of the court under section 622, the court shall proceed with the trial of the indictment and, where the organization is convicted, section 735 applies.

Sections 620 to 623 of the *Criminal* Code deal with appearances and notices. The heading prior to section 620 of the *Criminal Code*, together with sections 620 through 623, inclusive, are the subject of these Bill C-45 amendments. Section 620 deals with the filing of an appearance and the entering of a plea by counsel or agent. Section 621 had dealt with proper notice under the *Criminal Code* to corporations. Section 622 addressed the circumstance that a corporation does not appear in court in accordance with the notice provision of section 621. Section 623 had dealt with the appearance and entering of a plea by a corporation in a trial for a *Criminal Code* charge. The heading for these sections in the *Criminal Code* and their content are amended by replacing the word "corporation" with the new term "organization". These sections of the *Criminal Code* which relate to appearances, notice, default of appearance and the effect of a not guilty plea now apply to organizations rather than just corporations.

SECTION 12: REPLACES "CORPORATIONS" WITH "ORGANIZATION" WITH RESPECT TO PRESENCE IN COURT DURING TRIAL

12. Subsection 650(1) of the Act is replaced by the following:

650. (1) subject to subsections (1.1) and (2) and section 650.01, an accused, other that an organization, shall be present in court during the whole of his or her trial.

Section 650 of the *Criminal Code* establishes a general rule in criminal law that requires, unless the court orders otherwise, an accused to be present throughout the trial. Since a corporation is a legal fiction, there is no possible means by which a corporation can be compelled, other than through its counsel, agent or representative, to attend its full trial. Any corporation charged with a criminal offence is well advised to have a representative, apart from legal counsel, present at all times throughout the trial. Subsection 650(1) provided for an exception for corporations. Bill C-45 replaces the term "corporation", in the context of the exception, with the new term "organization". This section excludes an accused organization from the requirement that an accused must be present in court during a trial against the organization.

SECTION 13: REPLACES "CORPORATION" WITH "ORGANIZATION" WITH RESPECT TO SERVICE OF PROCESS

13. Section 703.2 of the Act is replaced by the following:

703.2 Where any summons, notice or other process is required to be or may be served on an organization, and no other method of service is provided, service may be effected by delivery

(a)   in the case of a municipality, to the mayor, warden, reeve or other chief officer of the municipality, or to the secretary, treasurer or clerk of the municipality; and

(b)   in the case of any other organization, to the manager, secretary or other executive officer of the organization or one of its branches.

Section 703.2 of the *Criminal Code* sets out the process for service of a summons, notice or other process required under the *Criminal Code* on a corporation. Service on a municipal corporation may be made to the mayor, warden, reeve or other chief officer of the corporation or to the secretary, treasurer or clerk of the corporation. Service on any other corporation may be to the manager, secretary or other executive officer of the corporation or branch of the corporation. Bill C-45 does not substantively change this provision. It simply replaces the word "corporation" with the new term "organization". Service of process remains the same with the exception that organization replaces corporation. In the private sector, a manager, secretary or other executive officer may accept service on behalf of the organization.

SECTION 14: FACTORS TO CONSIDER WHEN SENTENCING AN ORGANIZATION.

14. The Act is amended by adding the following after section 718.2:

Organizations

718.21 A court that imposes a sentence on an organization shall also take into consideration the following factors:

(*a*) any advantage realized by the organization as a result of the offence;

(*b*) the degree of planning involved in carrying out the offence and the duration and complexity of the offence;

(*c*) whether the organization has attempted to conceal its assets, or convert them, in order to show that it is not able to pay a fine or make restitution;

(*d*) the impact that the sentence would have on the economic viability of the organization and the continued employment of its employees;

(*e*) the cost to public authorities of the investigation and prosecution of the offence;

(*f*) any regulatory penalty imposed on the organization or one of its representatives in respect of the conduct that formed the basis of the offence;

(*g*) whether the organization was – or any of its representatives who were involved in the commission of the offence were – convicted of a similar offence or sanctioned by a regulatory body for similar conduct;

(*h*) any penalty imposed by the organization on a representative for their role in the commission of the offence;

(*i*) any restitution that the organization is ordered to make or any amount that the organization has paid to a victim of the offence; and

(*j*) any measures that the organization has taken to reduce the likelihood of it committing a subsequent offence.

Sentencing either an individual or an organization that has been convicted of a criminal offence is perhaps the most significant decision the criminal justice system is called upon to make for most offenders. Most criminal charges in Canada are resolved by way of the plea of guilty rather than a trial. Therefore, sentencing in most criminal cases is the most important issue since the issue of guilt or innocence has been resolved by way of a guilty plea, often stemming from a plea bargain agreement. Federal parliament assigns criminal offences varying degrees of culpability, seriousness, and importance by the severity of the penalty assigned to the offence. As a result, there is the wide range of circumstances and the degree of criminal activity may result in a conviction. After a conviction, the judge has to follow the requirements of the *Criminal Code* in sentencing the accused. Sentencing, however, largely remains a matter of judicial discretion, with only a limited number of criminal offences defining minimum penalties. Sentencing is primarily a discretionary process in the Canadian criminal justice system. Under the *Criminal Code*, judges are required to base sentencing on the overriding principle of proportionality. The *Criminal Code* states the principle of proportionality as follows: "A sentence must be proportionate to the gravity of the offence and the degree of responsibility of the offender."[7]

One of the fundamental purposes of sentencing is to contribute, along with crime prevention initiatives, to respect for the law and the maintenance of a just, peaceful and safe society.[8] Bill C-45 does not amend or change any existing sentencing principle in the *Criminal Code* or in criminal law jurisprudence. However, Bill C-45 establishes a new list of criteria for sentencing organizations in the *Criminal Code* that must be considered by the sentencing judge. Although many of the principles under new section 718.21 have been previously considered in cases dealing with sentencing principles, they are especially important in view of the new formula for organizational criminal liability.

Section 718 of the *Criminal Code* deals with the general purpose of sentencing an accused after a conviction. Bill C-45 adds section 718.21 to this sentencing provision in order to expand and to clarify factors that the court must consider when imposing a sentence on an organization. The use of the word "must" in section 718.21 makes consideration of these 10 new sentencing criteria for an organization mandatory. These new sentencing powers of the court are not limited to accused who are

---

[7] R.S.C. 1985, c. C-46.
[8] *Ibid.*, s. 718.

convicted of a breach of section 217.1, the new offence of OHS criminal negligence. They apply to all criminal offences with respect to which organizations may be charged and convicted.

These new sentencing factors reflect the trend in sentencing currently being used in prosecutions under OHS regulations. The leading case on sentencing in OHS prosecutions is the decision in *R. v. Cotton Felts Ltd.*[9] The court set out five factors in that case that were to be considered when sentencing:

1) the size of the company involved,
2) the scope of the economic activity in issue,
3) the extent of actual and potential harm to the public,
4) the maximum penalty prescribed by the statute, and
5) the need to enforce regulatory standards by deterrents.

The Court of Appeal for Ontario held that this last factor of deterrents was to be of paramount importance. While courts in most Canadian jurisdictions have acknowledged the applicability of *Cotton Felts,*[10] the five sentencing factors have been added to by court decisions in other jurisdictions. For example, in the Alberta Provincial Court decision of *R. v. Fiesta Party Rentals (1984) Ltd.,*[11] the court added the following as factors to be considered in imposing sentences: the special circumstances of the victim, the intent or degree of negligence involved, the extent of the accused's attempts to comply with the legislation, the element of risk involved in the activity, the remorse expressed by the accused, the record of the accused, and the economic impact of the fine upon the accused's business.

Under the Bill C-45 amendments ten additional factors have been set out that shall be considered by courts when sentencing organizations. Factor (*a*) is any advantage realized by the organization as a result of the offence. Presumably, this would include any money saved by failing to

---

[9]   (1982), 2 C.C.C. (3d) 287.
[10] See *R. v. Canadian Forest Products Ltd.* (1997), 24 C.E.L.R. (N.S.) 6 (B.C. Prov. Ct.); *R. v. Fiesta Party Rentals* (1984) Ltd., [2000] A.J. No. 1679 (Alta. Prov. Ct.); *R. v. A OK Holdings (c.o.b. Minute Muffler),* [2002] M.J. No. 266 (Man. Prov. Ct.); *R. v. Bayview-Wellington Homes (Port Union) Inc. (c.o.b. Bayview-Wellington Homes),* [2003] O.J. No. 1111 (Ont. Ct. Jus.); *R. v. McGuire* (1994), 147 N.B.R. (2d) 341 (N.B. Prov. Ct.); *R. v. Halifax Water Commission* (1993), 120 N.S.R. (2d) 398 (N.S. Prov. Ct.); *R .v. Etsell* (1993), 113 Nfld. & P.E.I.R 203 (Nfld. S.C.T.D.); *R. v. Bradley Air Services Ltd. (c.o.b. First Air),* [1999] N.W.T.J. No. 101 (N.W.T. Terr. Ct.).
[11] *Fiesta Party Rentals, ibid.*

invest in OHS legislative or regulatory compliance that could have prevented the offence from occurring. This may also include specific decisions to forgo or delay implementation of safety devices or OHS training with respect to dangerous equipment, machinery or processes because of the cost involved. Factor (*b*) requires the court to consider the degree of planning that went into the offence along with the duration and complexity of the offence. Here, a distinction can be made between those offences that are committed with relatively little planning, and those that require careful planning and execution on behalf of the organization. The new offence of OHS criminal negligence does not require proof of planning to result in a conviction. It only requires proof of negligence as set out in section 219 of the *Criminal Code*. It remains to be seen if a lack of planning could be an important sentencing consideration for the new offence of OHS criminal negligence. Factor (*c*) allows the court to consider any attempt taken by an organization to avoid paying a fine or restitution. Converting or concealing assets will not assist an organization to receive a reduced fine; in fact, it may lead to an increased penalty. The economic impact of a sentence on the organization, as well as the impact on the employment of employees, is considered under factor (*d*). In OHS regulatory prosecution cases, however, it has been determined that while this is a valid sentencing factor, it will not be given as much weight as other factors.[12] Factor (*e*) makes the cost of prosecution and investigation of the offence a consideration in sentencing. It is not clear if this is limited to police investigation and prosecution or if it may also apply to OHS regulatory investigations and prosecutions.

Factor (*f*) involves any regulatory penalty imposed arising out of events or conduct that also formed the basis of the health and safety criminal negligence offence. Factor (*g*) looks to previous convictions or regulatory sanctions for similar conduct, either of the organization or one of its representatives, as factors to be considered in sentencing. Evidence that the organization has a history of similar convictions or behaviour will likely result in an increased penalty. This factor is much broader than just looking at a prior criminal record. Factor (*h*) concerns whether the organization has imposed any penalty on a representative for his or her role in an offence. This may be a mitigating factor for the organization, when it has recognized the role of a representative in causing the offence and taking steps to prevent a recurrence. However, it

---

[12]   *Ibid.* at para. 28; *R. v. Tech-Corrosion Service Ltd.*, (1986), 43 Alta. L.R. (2d) 88 (Q.B.) at 92.

may also be an aggravating factor if the organization has unfairly disciplined or discharged a representative as a scapegoat for the organization's crime. Factor (*i*) requires the court to consider any restitution ordered or amount paid to the victim by the organization. In *Fiesta Party Rentals,* discussed above, the corporate defendant had made a $20,000 donation to a charity in the name of the victim of a workplace accident. The court held that although the defendant was not entitled to a dollar for dollar credit for such a charitable donation it was a mitigating factor to be taken into account with all of the other factors. Finally, factor (*j*), any measures taken by the organization to reduce the likelihood of committing subsequent offences, can be considered as evidence of how seriously the organization is in preventing a recurrence. A genuine display of efforts to improve the OHS regulatory compliance by, for example, developing an effective OHS management system and reduce the risk of further incidents will likely be a significant mitigating factor in determining an appropriate sentence.

SECTION 15: REPLACES "CORPORATION" WITH "ORGANIZATION" WITH RESPECT TO PROBATION REPORTS.

15. Subsection 721(1) of the Act is replaced by the following:

721. (1) Subject to regulations made under subsection (2), where an accused, other than an organization, pleads guilty to or is found guilty of an offence, a probation officer shall, if required to do so by a court, prepare and file with the court a report in writing relating to the accused for the purpose of assisting the court in imposing a sentence or in determining whether the accused should be discharged under section 730.

Section 721 of the *Criminal Code* is the statutory authorization for the preparation of a pre-sentence report by a probationary officer. The report prepared by a probation officer is an important aspect of the criminal justice system with respect to sentencing and follow up on offenders. The Lieutenant Governor in Council may make regulations prescribing the types of offences for which a court may require such a report. Subsection 721(3) delineates the mandatory required contents of a report by a probation officer. Subsection 721(1), before the Bill C-45 amendments, exempted corporations from the preparation of a report by a probation officer. This exemption continues for the new term "organization". However, it is not clear, given the broad additional optional conditions of probation dealt with in the 18 amendments section, whether or not there is any significance to this exemption for organizations. This section remains the same, with the exception that "corporation" is replaced with the new term "organization". An accused organization does not require a probation report, *per se*; however, reference should be made to new probationary powers in the section 18 amendments of this chapter.

SECTION 16: EVIDENCE OF PREVIOUS CONVICTIONS

16. Subsection 727(4) of the Act is replaced by the following:

(4) If, under section 623, the court proceeds with the trial of an organization that has not appeared and pleaded and convicts the organization, the court may, whether or not the organization was notified that a greater punishment would be sought by reason of a previous conviction, make inquiries and hear evidence with respect to previous convictions of the organization and, if any such conviction is proved, may impose a greater punishment by reason of that conviction .

Section 727 of the *Criminal Code* provided for notice to be given of intention to seek an increased sentence because of a prior conviction of both an individual and a corporation that is convicted of a criminal offence. If a criminal trial has proceeded *ex parte*, no greater punishment can be imposed unless the Crown prosecutor proves that it notified the accused of its intention to seek an increased sentence before the accused entered a plea prior to trial. This notice, commonly referred to as giving notice of greater punishment, now applies to both an individual and an organization under the Bill C-45 amendments. Evidence of previous convictions of the organization is allowed to be presented to and heard by the court even without notice to the organization that such evidence would be used. Greater punishment may be sought as a result of previous convictions. This provision of the *Criminal Code* overlaps with new section 718.21, which deals with the sentencing of organizations.

SECTION 17: ABSOLUTE AND CONDITIONAL DISCHARGES

17. Subsection 730(1) of the Act is replaced by the following:

730. (1) Where an accused, other than an organization, pleads guilty to or is found guilty of an offence, other than an offence for which a minimum punishment is prescribed by law or an offence punishable by imprisonment for fourteen years or for life, the court before which the accused appears may, if it considers it to be in the best interests of the accused and not contrary to the public interest, instead of convicting the accused, by order direct that the ac-

cused be discharged absolutely or on the conditions prescribed in a probation order made under subsection 731(2).

Section 730 of the *Criminal Code* provides a court with the power to order that an accused should not get a criminal record in connection with the offence with which they have been convicted. This provision, prior to Bill C-45, provided that if an accused other than a corporation is convicted of an offence for which no minimum punishment is prescribed, or for which the maximum punishment is less than fourteen years in jail, the court may, if it considers it to be in the best interest of the accused and not contrary to the general public interest, discharge the accused either absolutely or on conditions set by the court. An absolute discharge takes place immediately and the accused is deemed not to have been convicted or have a criminal record. A conditional discharge requires the accused to enter into a probation order for a specified period of time and does not come absolute until that time has passed. At that time, the accused will be deemed not to have been convicted. If the accused, however, breaches the terms of the probation order, he or she may be brought back before the court at which time a further and additional penalty may be imposed. All of these sentencing options would apply to an individual charged and convicted with the new OHS criminal negligence offence. However, the purpose of this amendment is to simply replace the word "corporation" with the new term "organization". An organization is still not eligible for a conditional or absolute discharge under section 718 of the *Criminal Code*.

SECTION 18: EXTENDS DEFINITION OF "OPTIONAL CONDITIONS" OF PROBATION.

18.(1). The definition "optional conditions" in subsection 732.1(1) of the Act is replaced by the following:

"optional conditions" means the conditions referred to in subsection (3) or (3.1)

Section 732.1 of the *Criminal Code* provides both compulsory and optional terms of probation. Subsection 732.1(2) sets out conditions of probation that the offender must always comply with; keep the peace and be of good behaviour, appear before the court when required to do so by order of the court, and notify the court or a probation officer in advance of any change of name or address, and promptly notify the court or the probation officer of any change of employment or occupation. In addition, optional conditions of probation are listed in subsection 732.2(3). Bill C-45 adds additional optional terms of probation in subsection 723.1(3.1). The definition of "optional conditions" is extended to include the conditions set out in subsection 3.1, which is added by section 18(2), listed below.

18.(2). Section 723.1 of the Act is amended by adding the following after subsection (3).

(3.1) The court may prescribe, as additional conditions of a probation order made in respect of an organization, that the offender do one or more of the following:

(*a*)   make restitution to a person for any loss or damage that they suffered as a result of the offence;

(*b*)   establish policies, standards and procedures to reduce the likelihood of the organization committing a subsequent offence;

(*c*)   communicate those policies, standards and procedures to its representatives;

(*d*)   report to the court on the implementation of those policies, standards and procedures;

(*e*)   identify the senior officer who is responsible for compliance with those policies, standards and procedures;

(*f*)   provide, in the manner specified by the court, the following information to the public, namely,

(i)   the offence of which the organization was convicted,

(ii)   the sentence imposed by the court, and

(iii) any measures that the organization is taking – including any policies, standards and procedures established under paragraph (b) – to reduce the likelihood of it committing a subsequent offence; and

(*g*)   comply with any other reasonable conditions that the court considers desirable to prevent the organization from committing subsequent offences or to remedy the harm caused by the offence.

(3.2) Before making an order under paragraph (3.1)(b), a court shall consider whether it would be more appropriate for another regulatory body to supervise the development or implementation of policies, standards and procedures referred to in that paragraph.

Bill C-45 adds new optional terms of probation specifically for organizations convicted of criminal offences. There was no predecessor provision in the *Criminal Code* dealing with special orders of probation for corporations. The result of this addition to the *Criminal Code* will be a higher level of post-conviction scrutiny of organizations by the courts, probation officers, and other regulatory bodies such as the applicable OHS regulator. This specific authorization for an OHS regulatory body to supervise the development or implementation of OHS policies, standards and procedures set out in these provisions, indicates an opportunity for an increased working relationship between the criminal justice system and the OHS regulators in the applicable jurisdiction. This change in the *Criminal Code* integrates the new OHS legal duty, the new offence of OHS criminal negligence, the new sentencing and probationary powers of the courts, with OHS regulators, as regulated by their statutes across Canada.

This new section gives courts special powers with respect to terms of probation just for organizations. These provisions would not apply to an individual who has been convicted of OHS criminal negligence. Probation orders may include restitution, establishing OHS policies, standards and procedures, communicating them to its representatives, reports on their implementation, identify the senior officer responsible for compliance and reporting to the court on the progress of improved occupational health and safety policies, standards and procedures. These are steps that are determined by the court to ensure there are no repeat offences, dangers in the workplace or workplace accidents.

Subsection 732.1 (3.1)(a) allows a court to order an organization to make restitution for any loss or damage that they suffered as a result of the offence. Remoteness of such loss or damage is not addressed, which raises the question of how remote damage must be before this section will not apply. Workers' compensation legislation provides a system to compensate injured workers and their dependents. This subsection raises questions of whether additional compensation may be part of a restitution order.

Subparagraph (b) of this new probationary provision of the *Criminal Code* allows the court to require an organization, after a conviction, to establish policies, standards and procedures to reduce the likelihood of the organization committing a subsequent offence. This probationary power of the courts is similar to current requirements under current OHS statutes and regulations across Canada. The Bill C-45 legal duty requires "reasonable steps" to be taken; however, if there is such a failure to take those reasonable steps and a conviction results, then the court is now authorized to impose policies, standards and procedures that ought to have been in place to reduce the likelihood of a recurrence of the accident, injury, or death in the workplace. Criminal courts do not have any previous authority or experience in establishing OHS policies, standards and procedures. The governmental body that does have authority and experience in regulating OHS policies, standards and procedures is the applicable OHS regulator. Therefore, it is very likely that this provision, if invoked as an appropriate probationary order, will be put in the hands of the applicable OHS regulator, rather than the police and probationary officers for ongoing scrutiny.

Subparagraph (c) gives a court the authority to order that the OHS policies, standards and procedures that have been required be communicated to its representatives. Communication is a critical component to an effective OHS management system. Therefore, this probationary order appears to be consistent with the requirement for an improvement in the OHS management system of the organization that has been convicted to ensure that a recurrence of the offence is prevented.

Subparagraph (d) requires the organization to report to the court on the implementation of those policies, standards and procedures. This heightened accountability for the organization to report its improvement to the OHS management system complements the new above-mentioned probationary orders. Further, it holds the organization accountable, to a criminal court, for the improvement to its OHS management system by legislating and communicating new OHS policies, standards and procedures. Further, subparagraph (e) allows the court to identify a senior officer of the organization who is responsible for compliance with those policies, standards and procedures. This individual, presumably, would be the individual reporting to the court on the implementation of the improved OHS management system.

Subparagraph (f) is novel and may be controversial. It allows a court to compel the organization to provide information to the public regarding the offence with which the organization was convicted, the sentence imposed by the court, and any measures that the organization is taking to reduce the likelihood of committing a subsequent offence. In practical terms, this may require an organization to take out an ad in a national newspaper identifying that it has been charged and convicted with the offence of OHS criminal negligence. The organization will also be required to identify the nature of the sentence imposed on the organization, and any individual employed by the organization, as the case may be. Further, the organization may also be required to tell the public, in this national advertisement, that it has been compelled by a probationary order to take certain steps to improve its OHS policies, standards and procedures. These types of probationary orders, known as public shaming orders, have become increasingly popular in the United States. Public shaming as a form of punishment has its historical roots in the middle ages in Europe when individuals convicted of crimes were publicly flogged, placed in public stocks, and made the subject of very public hangings.

Subparagraph (g) also gives the court a broad power to compel the organization to comply with any other reasonable conditions that the court considers desirable to prevent the organization from committing the offence again or to remedy the harm caused by the offence. This would permit the court to order the organization to comply with OHS statutes and regulations, provide OHS training to managers, supervisors and workers, and to ensure that an effective OHS management system was in place. These further powers of the court, under the probationary powers of the *Criminal Code*, are broader than the provisions of any provincial or federal OHS statute. Although OHS statutes and regulations in Canada are increasingly prescriptive with respect to the means by which they require employers to ensure the health and safety of workers, they have generally stopped short of giving courts the extensive probationary powers that are in this new provision of the *Criminal Code*. A more complete review and understanding of Canadian OHS law is offered in chapter 5 of this book.

Subparagraph 732.1(3.2) gives a court the power to consider the most appropriate regulatory body to supervise the development and implementation of the policies, standards and procedures referred to in the previous subparagraph. In other words, a court may reasonably consider the role and authority of the applicable OHS regulator, as established by the applicable OHS statute, to supervise this aspect of the new probationary powers of the court. This provision clearly indicates the need for a close working relationship between the applicable OHS regulator, the police, the crown attorney, and court in the prosecution, sentencing and probationary orders relating to the new offence of OHS criminal negligence.

SECTION 19: REPLACES "CORPORATION" WITH "ORGANIZATION" WITH RESPECT TO POWERS TO IMPOSE FINES

19. The portion of subsection 734(1) of the Act before paragraph (a) is replaced by the following:

734. (1) Subject to subsection (2), a court that convicts a person, other than an organization, of an offence may fine the offender by making an order under section 734.1

Section 734 of the *Criminal Code* provides authority to a court to impose a fine in lieu of the term of imprisonment, or in addition to another punishment, as the court sees fit. If there is a minimum term of imprisonment, the court may not substitute the minimum term of imprisonment with a fine. Before imposing the fine the court must be satisfied that the convicted offender is able to pay the fine or discharge the payment of fine through a fine option program. This power given to the court explicitly exempted corporations. Although Bill C-45 replaces the word "corporation" with the new term "organization" in this section, the exemption has been continued for organizations. The court does not have the power to impose a fine on an organization under subsection 734(1) of the *Criminal Code*, in lieu of imprisonment. Like a corporation, an organization may not be imprisoned.

SECTION 20: REPLACES "CORPORATION" WITH "ORGANIZATION" AND INCREASES LIMIT OF POSSIBLE FINES

20. (1) The portion of subsection 735(1) of the Act before paragraph (a) is replaced by the following:

> 735. (1) An organization that is convicted of an offence is liable, in lieu of any imprisonment that is prescribed as punishment for that offence, to be fined in an amount, except where otherwise provided by law,

Subsection 735(1) of the *Criminal Code* previously addressed the maximum fines for corporations in lieu of any provision for imprisonment. The appropriateness of fines against corporations was addressed in the prosecution of *R. v. McNamara* (no. 2).[13] In that case, the court upheld fines imposed on corporations for conspiracy to defraud various government agencies in the rigging of various bids on dredging contracts. The fines for convictions of those indictable offences ranged from $50,000 to $2 million. Factors which the court considered included specific and general deterrence. The court also held that it was of the view that it could appropriately consider profits and anticipated profits determining the appropriate size of the fine given the nature of the crime. Bill C-45 replaces the term "corporation", with the new term "organization", with respect to this provision that deals with the availability of fines for organizations convicted of an offence. Subsection 735(1) is also the authority for the proposition that there is no maximum fine set for an organization convicted of an indictable offence under the *Criminal Code*.

---

[13]   (1981), 56 C.C.C. (2d) 516 (Ont. C.A.).

20. (2). Paragraph 735(1)(b) of the Act is replaced by the following:

(b) not exceeding one hundred thousand dollars, where the offence is a summary conviction offence.

Subparagraph 735(1)(b) of the *Criminal Code* is amended to increase the maximum fines against organizations from $25,000 to $100,000 for summary conviction offences. Summary conviction offences are the less serious means of prosecuting a criminal offence. This is a significant increase and will have to be taken seriously by organizations. Although this is a 400% increase in the fines for organizations convicted of summary conviction offences, it still falls short of a number of penalties available under OHS statutes across Canada. Chapter 5 of this book deals with Canadian OHS laws and their penalties in more detail, including the maximum penalties available.

20. (3). Subsection 735(2) of the Act is replaced by the following:

(2) Section 734.6 applies, with any modifications that are required, when an organization fails to pay the fine in accordance with the terms of the order.

Subsection 735(2) of the *Criminal* Code deals with the enforcement of unpaid fines. The process remains the same, with the exception that the new term "organization" replaces a corporate offender. If an organization is in default of payment of a fine or forfeiture, the order imposing payment may be filed and entered as a judgment for that amount against the organization. A judgment of the court may be enforced by a variety of means to enforce both criminal and civil orders of the courts.

SECTION 21: REPLACES "CORPORATION" WITH "ORGANIZATION" WITH RESPECT TO APPEARANCES FOR SUMMARY CONVICTION OFFENCES

21. Subsection 800(3) of the Act is replaced by the following:

(3) Where the defendant is an organization, it shall appear by counsel or agent and, if it does not appear, the summary conviction court may, on proof of service of the summons, proceed *ex parte* to hold the trial.

Section 800 of the *Criminal Code* provides that where both a defendant and prosecutor appear, the summary conviction trial should commence. The defendant may appear in a summary conviction prosecution by counsel or agent, but the court may require the defendant's personal appearance and enforce this requirement through a warrant. This subsection remains the same, with the exception that the new term "organization" replaces corporation. If an organization is charged with a summary conviction offence and does not appear, the court may proceed *ex parte* with the trial.

SECTION 22: COORDINATING AMENDMENT

22. On the later of the coming into force of section 9 of this Act and section 34 of the *Criminal Law Amendment Act, 2001*, section 556 of the *Criminal Code* is replaced by the following:

556. (1) An accused organization shall appear by counsel or agent.

(2) Where an accused organization does not appear pursuant to a summons and service of the summons on the organization is proved, the provincial court judge or, in Nunavut, the judge of the Nunavut Court of Justice

(*a*) may, if the charge is one over which the judge has absolute jurisdiction, proceed with the trial of the charge in the absence of the accused organization; and

(*b*) shall, if the charge is not one over which the judge has absolute jurisdiction, fix the date for the trial or the date on which the accused organization must appear in the trial court to have that date fixed.

(3) If an accused organization appears and a preliminary inquiry is not requested under subsection 536(4), the provincial court judge shall fix the date for the trial or the date on which the organization must appear in the trial court to have that date fixed.

(4) If an accused organization appears and a preliminary inquiry is not requested under subsection 536.1(3), the justice of the peace or the judge of the Nunavut Court of Justice shall fix the date for the trial or the date on which the organization must appear in the trial court to have that date fixed.

Section 556 of the *Criminal Code* has been amended to replace corporation with the new term "organization". This section coordinates with a previous amendment to this section which sets out the effects of non-appearance as well as what happens when a preliminary inquiry is not requested.

SECTION 23: COMING INTO FORCE DATE

23. The provisions of this Act, other than section 22, come into force on a day or days to be fixed by order of the Governor in Council.

This section of Bill C-45 proscribes when the changes to the *Criminal Code* will come into force. Bill C-45 received first reading on June 12, 2003 in the House of Commons. The legislative process ended with the

CHAPTER 4: DETAILED ANALYSIS OF BILL C-45

third reading in the Senate on October 30, 2003 and Royal Assent given on November 7, 2003. Bill C-45, however, did not come into force when it received Royal Assent. As section 23 of Bill C-45 clearly states, this Act will come into force on a day to be fixed by order of the Governor in Council. The federal cabinet, acting as the Governor in Council, fixed the date that Bill C-45 came into force to beMarch 31, 2004. Effective that day, all the provisions of Bill C-45 took effect, including section 217.1, the new OHS duty. The Bill C-45 amendments are prospect and not retroactive. They only apply from the date they came into force as determined by the order of the Governor in Council.

CHAPTER 5

UNDERSTANDING CANADIAN OCCUPATIONAL HEALTH AND SAFETY LAW

INTRODUCTION TO CANADIAN OCCUPATIONAL HEALTH AND SAFETY LAW

Canadian occupational health and safety law has developed over the last 50 years as a reaction to workplace accidents, injury and death. Canadian workers' compensation legislation predates Canadian OHS law. Since the Meredith report of 1913, a no-fault system provided compensation to workers injured on the job. It appears that governments, employers, unions, and other workplace stakeholders at the beginning of the 20th century were more concerned with providing workers' compensation protection for injured employees and protecting employers from lawsuits than from preventing accidents in the workplace.[1] Canadian OHS law has developed significantly in the second half of the 20th century.

Workplace health and safety has no specific jurisdictional designation under the Canadian *Constitution Act, 1867*. The *Constitution Act, 1867* sets out a division of powers between the federal and provincial governments. OHS is not the subject of an explicit reference to the division of powers between the federal and provincial governments. Therefore, courts have been called upon to determine whether the provincial or the federal government has authority to regulate the workplace by OHS legislation. Approximately 10 per cent of Canadian workplaces are federally regulated and 90 per cent are provincially regulated for the purposes of labour relations, employment standards,

---

[1] For a more complete introduction to the origins of this subject see N.A. Keith, *Canadian Health and Safety Law* (Aurora: Canada Law Book, 2003) at 1:10.

workers' compensation and OHS. Therefore, the vast majority of Canadian workers are regulated by provincial OHS statutes and regulations.

OHS statutes generally set the framework for the health and safety requirements, standards and procedures in the jurisdiction in which they apply. OHS statutes in Canada are based on the internal responsibility system, which is an overlapping system of rights and responsibility of workplace stakeholders. A detailed discussion of the internal responsibility system is in the second section of this chapter. Canadian OHS law is also based on the external responsibility system, the lawful authority establishing government regulatory accountability. The external responsibility system has two means of enforcement of OHS requirements, standards and procedures. First is the issuance of orders or directions by inspectors or officers, employed by various government regulators. The issuance of an order may be to stop working immediately, or to change a work practice within a reasonable period of time.

Second is the laying of charges under OHS laws in Canada as a means of enforcing the duties for various workplace parties. It is an OHS regulatory offence to contravene these duties. This approach to establishing an offence is different than the establishment of a crime under the *Criminal Code*. Under the *Criminal Code* certain conduct is expressly designated to be a criminal offence.

Canadian OHS law provides for enforcement, in part, by prosecutions brought as quasi-criminal, strict liability offences. There are three types of offences known to Canadian law identified by the Supreme Court of Canada in *R. v. Sault Ste. Marie* (City):[2]

(i)   *mens rea* offences;
(ii)   strict liability offences;
(iii) absolute liability offences.

The enforcement of Canadian OHS law against workplace stakeholders that have legal duties has been a growing trend across Canada, and the incidence of enforcement by way of prosecution has increased since the mid-1970s. Workplace stakeholders may include the employer, supervisors, officers, directors, professional engineers, architects, suppliers, workers and others. Although many workplace

---

[2]   *R. v. Sault Ste. Marie (City)* (1978), 85 D.L.R. (3d) 161.

stakeholders have legal duties under Canadian OHS law, employers are the primary stakeholders that are prosecuted with OHS offences.

The establishment of legal duties on workplace parties for OHS is not new. The common law has long placed a duty on an employer to provide a safe workplace for workers free of unnecessary and unreasonable hazards. Under the Anglo-Canadian common law, if an employer failed to meet a reasonable standard of health and safety for workers in the workplace and an injury resulted, the employer could be successfully sued for negligence. However, with the development of workers' compensation legislation, the legal right of a worker to sue an employer for breach of this common law duty was effectively, and almost universally, terminated. There are still some limited circumstances in which third party lawsuits may be available for the injured party. Bill C-45 recognizes that public safety is also an important part of workers' safety. A member of the public who is not a worker has a full right to sue for negligence if injured by a workplace accident.

Under Canadian OHS law, workplace stakeholders and employers must meet certain prescribed legislative duties and responsibilities to protect employees and workers. Some Canadian OHS statutes focus on employees only, while others apply to the broader term "workers". For example, Part II of the *Canada Labour Code* for federally regulated employers emphasizes the duty to protect employees only. On the other hand, the Ontario *Occupational Health and Safety Act*[3] duties on workplace stakeholders relate to the protection of workers, not just direct employees. Therefore, Bill C-45's new duty in section 217.1 of the *Criminal Code* goes farther than any current OHS statute since it addresses public as well as worker safety.

Bill C-45 has introduced an explicit new legal duty regarding workplace health and safety within the criminal negligence sections of the *Criminal Code*. This legal duty theoretically establishes one standard for OHS duties consistent across the country. The legal duty provides that those who undertake or have the authority to direct how another person does work or performs a task are required to take reasonable steps to prevent bodily harm to any person arising from the work. The framers of the legislation clearly intend to impose this duty on a wide range of personnel in the workplace. However, the duty to take reasonable steps in section 217.1 is not defined. All managerial personnel — regardless of whether they are front-line supervisors or senior officers who have the

---

[3]   R.S.O. 1990, c. O.1.

power to direct work — would be responsible for ensuring that all reasonable steps are taken to prevent bodily harm.

The duty is not confined to managers, supervisors, and officers or possibly directors; any individual employee who undertakes to direct another how to perform any task in the workplace would also be subject to the duty. As attractive as this broad duty may be from a policy perspective, it is not without its problems. Imposing this obligation on employees who direct how another worker performs a task may have the potential to create a chilling effect on working supervisors or lead hands, who may become reluctant to assume such roles if they attract the potential risk of criminal liability for even minor direction or supervision of others.

The legal duty requires that those subject to the duty take reasonable steps to prevent bodily harm to any person. Therefore, there would exist an obligation to workers and to the public at large. The public would include visitors to or volunteers in the workplace. This part of the duty will also be of critical importance in settings where the public is in physical attendance where the work is performed, or where the public could be affected by adverse consequences arising from work activities.

Canadian OHS law is enforced predominantly by issuing orders and by the prosecution of employers. Under Bill C-45, it is not clear who in the workplace would most likely be prosecuted for breach of the duty. Section 217.1 established a blanket criminal offence, applying equally to all who direct the workplace activities of others. Unlike Canadian OHS law that currently exists, there appears to have been no effort on the part of the drafters of Bill C-45 to clarify the nature or hierarchy of the duty owed. Instead, the Bill effectively states a broad and far-reaching duty, with no corresponding definitions, parameters, or guidelines as to how it will apply to various organizational decision-makers. If the current OHS prosecutions are any indication of how OHS criminal negligence charges will be laid, then there will likely be a much higher incidence of employers prosecuted than supervisors and more supervisors charged than workers. Workers sometimes fail to comply as do supervisors and senior management. However, government OHS regulators appear to predominantly focus on supervisors, senior management, and employers in laying charges under applicable OHS legislation. Time will tell if the same will apply to the new offence of OHS criminal negligence.

THE INTERNAL RESPONSIBILITY SYSTEM

As discussed briefly above, the internal responsibility system is the underlying concept and philosophy behind most modern developments in Canadian OHS law. The internal responsibility system suggests that it is the workplace stakeholders who are best able to identify, assess, and either eliminate or control hazards in the workplace. Modern Canadian OHS law establishes requirements, standards and procedures that the workplace stakeholders must comply with. The internal responsibility system recognizes the limits that governments at all levels have to provide full inspection, scrutiny, and enforcement of OHS statutes and regulations in every workplace in Canada. The system is manifest by a series of overlapping legal duties, rights, and responsibilities in OHS statutes and regulations.

The internal responsibility system has been described by Professor Katherine Swinton as follows:

> The phrase "internal responsibility system" was coined by Professor James Ham in his 1976 report on Occupational Health & Safety in Mines. Ham envisioned two types of responsibility systems: direct and contributively. The direct responsibility system would require management to define clear standards of work, assign responsibility for particular tasks, and then establish lines of accountability to ensure proper performance. This system of "direct responsibility" would be facilitated by a contribution system existing of an external auditing function, carried out by worker auditors and joint labour-management health and safety committees. A major flaw in regulating occupational health and safety in the past was, in Ham's opinion, the lack of worker participation.[4]

The internal responsibility system is made up of a number of different elements which may vary from jurisdiction to jurisdiction across Canada. However, the core elements of the internal responsibility system include the establishment of legal duties on various workplace stakeholders. Some jurisdictions, like Ontario, enumerate a long list of stakeholders that have legal duties including constructors, employers, supervisors, workers, licensees, professional engineers, architects, directors, officers and suppliers. Others, like the federal *Canada Labour Code*, Part II, only place legal duties on employers and employees. Regardless of the list of workplace stakeholders with duties, the overriding purpose of the legal duties are to ensure the health and safety of workers in the workplace.

---

[4]   K. Swinton, "Enforcement of Occupational Health and Safety Legislation: The Role of the Internal Responsibility System", in P. Swan and E. Swinton, eds., *Studies in Labour Law* (Toronto: Butterworths, 1983) at 143.

The internal responsibility also is demonstrated by the establishment of health and safety representatives and health and safety committees. Canadian workplaces with five or more workers, and less than twenty workers, generally require health and safety representatives. The exceptions are Alberta and Quebec where representatives are not required, and under the federal *Canada Labour Code*, Part II, where every workplace, even one with only one worker, must have a health and safety representative. These are worker, non-management members of the workplace who are designated to represent the interests of workers with respect to health and safety. They may inspect the workplace, request information from their employer about workplace hazards, and investigate workplace accidents. Health and safety committees are required in most jurisdictions, except Alberta and Quebec, where there are twenty or more workers in a particular workplace location. Committees may be required in Alberta or Quebec based on the Minister's order or the type of industry. Health and safety committees are made up of management and worker members, usually in equal number. Health and safety committees have the legislative right to meet, inspect the workplace, make inquiries of employers, and participate when the regulator investigates the workplace. In many jurisdictions, health and safety committees must be consulted before new OHS programs are introduced.

There are jurisdictions across Canada that regulate toxic substances and controlled products through various regulatory regimes. The purpose is to set standards that employers must follow to ensure that workers are not exposed to biological, chemical, physical, and ergonomic hazards. The Workplace Hazardous Material Information System, also known as WHMIS, is a national, legislative regime that has established hazard identification, assessment, and safe handling measures for controlled products. This system provides information to workers by way of mandatory training, labelling containers of control products, and providing material safety data sheets with additional information on how to avoid exposure to the control products and what to do in the event of exposure. WHMIS is a rare Canadian example where federal and provincial governments cooperated on the subject of OHS workers, resulting in a nationally consistent program to regulate and provide training with respect to controlled products. WHMIS training in one part of the country will be valid in another part of the country. The system is an example of the internal responsibility system at its national consistent best.[5]

Although the internal responsibility system is discussed by legislators and policy makers when amending various OHS statutes and regulations across Canada, for a long time there was no legal definition of the internal responsibility system. Although it was often referred to, it was never legally defined. During the Westray Inquiry, there was a lot of discussion about the need to improve OHS legislation and the internal responsibility system. The internal responsibility system was both lauded and criticized, but particularly criticized by the United Steelworkers Union of America at the Westray Inquiry.

In the post-Westray disaster era, Nova Scotia made significant changes to its OHS statute, one of which was providing a definition of the internal responsibility system. In section 2 of the Nova Scotia OHS statute, the following definition of the internal responsibility system is provided:

> The foundation of this Act is the Internal Responsibility System which
>
> (*a*) is based on the principle that
>
> > (i)   employers, contractors, constructors, employees, and self-employed persons at a workplace, and
> >
> > (ii) the owner of a workplace, a supplier of goods or provider of an occupational health or safety service to a workplace or an architect or professional engineer, all of whom can affect the health and safety of persons at the workplace,
>
> share the responsibility for the health and safety of persons at the workplace;
>
> (*b*) assumes that the primary responsibility for creating and maintaining a safe and healthy workplace should be that of each of these parties, to the extent of each party's authority and ability to do so;
>
> (*c*) includes a framework for participation, transfer of information and refusal of unsafe work, all of which are necessary for the parties to carry out their responsibilities pursuant to this Act and the regulations; and
>
> (*d*) is supplemented by the role of the Occupational Health and Safety Division of the Department of Labour, which is not to assume responsibility for creating and maintaining safe and healthy workplaces, but to establish and clarify the responsibilities of the parties under the law, to support them in carrying out their responsibilities and to intervene appropriately when those responsibilities are not carried out.

---

[5]   For further information on the workplace hazardous material information system see N.A. Keith, *Canadian Health and Safety Law* (Aurora: Canada Law Book, 2003) at c. 4.

In summary, the internal responsibility system is the foundational concept of Canadian OHS law. It provides legal requirements, standards and procedures for workplace health and safety, promotes self regulation of workplace hazards, and reduces the need for government regulators to intervene in the workplace. Ultimately and practically, employers have the highest level of responsibility under the internal responsibility system.

OHS LAW GENERAL DUTY CLAUSES

Part of the internal responsibility system, discussed above, has been the establishment of various duties on workplace stakeholders. Employers and senior management have many legal duties to ensure that a workplace is healthy and safe for employees and workers generally. In addition to specific duties, and regulations setting out control measures for employees, Canadian OHS statutes also have provisions known as general duty clauses. These general duty clauses provide a very broadly worded statement requiring employers, and on occasion other parties, to take all reasonable precautions for the health and safety and protection of workers in the workplace. General duty clauses have similarities to new section 217.1 of the *Criminal Code* introduced by Bill C-45. Since there is this similarity, they will be briefly reviewed here.

It is likely that judicial interpretation of the new offence established by Bill C-45, OHS criminal negligence, will refer to the statutory provisions, case law and judicial interpretation of the already-existing general duty clauses in health and safety legislation. Employers are the most frequently charged parties for violations of OHS statutes, and a large number of these charges relate to an alleged breach of the general duty to provide a safe workplace.[6]

The general duty clauses in various Canadian jurisdictions are quite similar. General duty clauses place broad health and safety responsibility on employers in their applicable jurisdictions. There is a relatively consistent pattern in the language of general duty clauses across various Canadian jurisdictions. To provide some review and analysis, the current general duty clauses are reviewed below.

---

[6] R.S.O. 1990, c. O.1, s. 25(1)(a), (b) and (d). In addition to general duty clauses, employers also have a specific duty with respect to safety requirements under other statutes such as the *Building Code Act*, 1992, S.O. 1992, c. 23. Specifically, an employer should obtain the opinion of a professional engineer or building code expert with respect to this obligation under the Ontario health and safety statute.

The federal jurisdiction, regulated by the *Canada Labour Code*, Part II,[7] provides employers with a very broad duty to ensure the health and safety of employees. It is interesting to note that the *Canada Labour Code*, Part II does not broadly protect workers, as is the case with most applicable OHS statutes across Canada, but rather restricts the employer's general duty clause to that of employees. Bill C-45, as discussed above, protects all individuals in the workplace, including workers and members of the public, who may be at risk from hazards or activity. The federal legislation general duty clause states, "Every employer shall ensure that the health and safety at work of every person employed by the employer is protected."[8]

In British Columbia, unlike other Canadian jurisdictions, the legislature establishes its employer and other stakeholder duties under Part III of the *Workers Compensation Act*[9] as applied by way of the *Occupational Health and Safety Regulation*.[10] The general duty clause under the *Workers Compensation Act* states: "Every employer must ensure the health and safety of all workers working for that employer."[11]

In Alberta, the *Occupational Health and Safety Act*[12] provides the following general duty clause: "2(1) Every employer shall ensure, as far as it is reasonably practicable for the employer to do so, (a) the health and safety of (i) workers engaged in the work of that employer...". Arguably, the Alberta general duty clause is the least broad and expansive of the general duty clauses in Canadian health and safety law. Further, the Alberta provision moderates the extent and potential breadth of application of the general duty clause by the phrase "as far as it is reasonably practicable for the employer to do so".

Saskatchewan has adopted a general duty clause similar to that of Alberta. The Saskatchewan *Occupational Health and Safety Act, 1993*[13] states:

3. Every employer shall:

(a) ensure, insofar as is reasonably practicable, the health, safety and welfare at work of all of the employer's workers.

---

[7] R.S.C. 1985, c. L-2.
[8] *Ibid.*, s. 124.
[9] R.S.B.C. 1996, c. 492.
[10] B.C. Reg. 296/97.
[11] R.S.B.C. 1996, c. 492, s. 115(1).
[12] R.S.A. 2000, c. O-2.
[13] R.S.S. 1993, c. O-1.1.

In that sense, in Manitoba, the legislature adopted similar language in its general duty clause. The Manitoba *Workplace Safety and Health Act*[14] states:

4(1) Every employer shall in accordance with the objects and purposes of this Act

    (*a*) ensure, so far as is reasonably practicable, the safety, health and welfare at work of all his workers...

The Ontario *Occupational Health and Safety Act*[15] establishes the general rights and responsibilities of government, employers and workers, and sets minimum health and safety standards for the workplace. With the duty to take every reasonable precaution under the Ontario health and safety statute, employers have a general duty to "take every precaution reasonable in the circumstances for the protection of a worker."[16] This duty may require an employer to take precautions that include the development of an OHS management system and the suspension or discharge of workers for unsafe work practices.[17] Supervisors must also take every precaution reasonable in the circumstances for the protection of a worker in a similar general duty.[18]

In Quebec, the health and safety legislation states: "Every employer must take the necessary measures to protect the health and ensure the safety and physical well-being of his worker." [19]

In New Brunswick, the *Occupational Health and Safety Act*[20] requires that:

9(1) Every employer shall

    (*a*) take every reasonable precaution to ensure the health and safety of his employees ...

As such, "foresee ability" is a critical factor in the due diligence standard. In New Brunswick, like other provinces, an OHS program must include activities designed to prevent the recurrence of accidents — analyzing jobs and work procedures to identify hazards and taking steps to eliminate or reduce those hazards. This concurs with the Province of

---

[14] C.C.S.M. c. W210.
[15] R.S.O. 1990, c. O.1.
[16] *Ibid.*, s. 25(2)(h).
[17] Ministry of Labour Legal Branch, *Interpretation Opinions*, (Ottawa: Ministry of Labour, November 1983) at 9; cited in N.A. Keith, *Canadian Health and Safety Law*, Looseleaf ed., (Aurora: Canada Law Book, 1997) at 3:40.2(6)(1).
[18] R.S.O. 1990, c. O.1, s. 27(2)(c).
[19] *An Act respecting Occupational Health and Safety*, R.S.Q., c. S-2.1, s. 51.
[20] S.N.B. 1983, c. O-0.2.

Prince Edward Island's *Occupational Health and Safety Act,*[21] where the general duty clause reads as follows:

> 13(1) Every employer shall take every precaution that is reasonable in the circumstances to
>
>> (*a*) ensure the health and safety of persons at or near the workplace;

To the same effect, in Nova Scotia, the *Occupational Health and Safety Act,*[22] amended substantially after the Westray mine Disaster, states that:

> 13(1) Every employer shall take every precaution that is reasonable in the circumstances to
>
>> (a) ensure the health and safety of persons at or near the workplace;

In the Province of Newfoundland, the *Occupational Health and Safety Act,*[23] with more gender inclusive language, states as follows: "An employer shall ensure, where it is reasonably practicable, the health, safety and welfare of his or her workers."

In the Northwest Territories and Nunavut, the *Safety Act*[24] has an interesting and rather lengthy general duty clause, which states as follows:

> 4. Every employer shall
>
>> (*b*) take all reasonable precautions and adopt and carry out all reasonable techniques and procedures to ensure the health and safety of every person in his or her establishment; ...

Finally, the Yukon Territory would appear to be the only Canadian jurisdiction without a general duty clause. Although the Yukon *Occupational Health and Safety* Act[25] places duties on employers with respect to ensuring that workplace machinery, equipment and processes are safe, and that work techniques and procedures are used to reduce the risk of occupational illness and injury, as well as other duties relating to instruction, hazard awareness and general compliance with the Act, there is no apparent general duty clause in the Yukon Territory statute.

The OHS general duty clauses are somewhat similar in their language to new section 217.1 of the *Criminal Code* introduced by Bill C-45. The phrase "reasonable steps" in section 217.1 is arguably not as broad or

---

[21] R.S.P.E.I. 1988, c. O-1.
[22] S.N.S. 1996, c. 7.
[23] R.S.N.L. 1990, c. O-3, s. 4.
[24] R.S.N.W.T. 1988, c. S-1.
[25] R.S.Y. 2002, c. 159.

strict as language that uses words such as "all", "every", and "ensure" found in many of the general duty clauses. The requirement to take "reasonable steps" is similar to the phrases "reasonable precautions" and "so far as it is reasonably practicable". Therefore, whether or not section 217.1 of the *Criminal Code* was specifically intended to be similar to a general duty clause, its purpose is clearly the same, to ensure the health and safety of workers and that the public are protected from bodily harm.

ENFORCEMENT OF OHS LAW AND DUE DILIGENCE

Canadian OHS law is made up of statutes, regulations and industry codes and practices that are often adopted or incorporated by reference into the statutes or the regulations. The goal, generally speaking, of OHS law is to prevent workplace accidents, injury, and death. It also holds workplace stakeholders responsible for failure to comply with OHS law. Canadian OHS statutes focus on protecting workers and not specifically members of the public. Bill C-45, on the other hand, focuses equally on both worker and public safety. OHS law emphasizes the role of workplace stakeholders in taking responsibility to identify, assess, and control workplace hazards, while Bill C-45 requires compliance with the new legal duty under section 217.1, failing which an individual or an organization may face prosecution for the offence of OHS criminal negligence.

The enforcement of OHS law is primarily by two means. Government regulators have the authority to issue orders or directions to comply with the OHS statutes and regulations. If a workplace stakeholder with a duty fails to comply with that duty, the OHS regulator may issue an order or direction requiring immediate compliance, or compliance within a reasonable period of time. This is the first means of enforcement of OHS law in Canada. Workplace stakeholders are given the right to appeal such orders or directions. Failure to comply with an order or direction, without commencing an appeal, is an offence.

The second means of enforcement is by way of quasi-criminal prosecution of OHS regulatory offences. These are not criminal offences but are similar in that they result in a government prosecution. Bill C-45, on the other hand, establishes a true criminal offence for breach of the new OHS duty under section 217.1 of the *Criminal Code*. Although the enforcement of OHS law in Canada by way of prosecution is similar in its process to a criminal prosecution, the legal characterization of a criminal charge is different from that of a strict liability OHS regulatory offence.

OHS statutes across Canada establish OHS regulatory offences. These offences, through a strict liability, give rise to the defence of due diligence. The classification of offences and the defence of due diligence are discussed below.

The Supreme Court of Canada, in *R. v. Sault Ste. Marie*,[26] indicated that strict liability offences, although not true criminal offences, are quasi-criminal in nature. Provincial and federal health and safety offences have also been defined as regulatory public welfare offences designed for the protection of public worker and interest. Provincial legislatures and the federal parliament have the power to create either true crimes or *mens rea* offences, strict liability or absolute liability offences. Although the criminal law power is constitutionally assigned to the federal government, a provincial offence may be classified as a true crime or *mens rea* offence. These three types of offences have been introduced early in this chapter. However, a more complete explanation is needed to understand the types of offences and the legal defence of due diligence.

(a) True Crimes or *Mens rea* offences: in *mens rea* offences, the prosecution must prove beyond a reasonable doubt the prohibited act and, either as an inference from the nature of the act committed or by additional evidence, the positive state of mind on the part of the accused, such as intent, knowledge or recklessness. In a criminal offence, the prosecution has the onus of proof throughout the trial. The onus of proof never shifts to the accused.

(b) Strict liability offences: in strict liability offences, as in absolute liability offences, the prosecution need only prove beyond a reasonable doubt that the defendant committed the prohibited act; the prosecution need not prove a fault element; and thereafter, the accused has the defence that it reasonably believed in a mistaken set of facts that, if true, would render the act or omission innocent, or that it has taken reasonable precautions to achieve compliance; these are the two branches of the due diligence defence. Most OHS offences are strict liability offences.

(c) Absolute liability offences: in absolute liability offences, the prosecution need only prove beyond a reasonable doubt that the accused committed the prohibited act, constituting the *actus reus* of the offence. There is no relevant mental element and it is no

---

[26] (1978), 85 D.L.R. (3d) 161.

due diligence defence that the accused was entirely without fault. More proof of the prohibited act will lead to a conviction.

Under the *R. v. Sault Ste. Marie* decision, the Supreme Court of Canada held that in a strict liability offence, the onus of proof shifts to the defendant to establish the defence of due diligence. The shifting of the onus of proof was not seen as unfair since the defendant alone had knowledge of what was done to avoid the commission of the prohibited act. It is therefore expected that the defendant would advance the defence of due diligence if it was available. There are two separate branches of the due diligence defence:

(i)   in the first branch, the defendant must prove that it reasonably believed in a mistaken set of facts which, if true, would render the prohibited act or omission innocent.

(ii) in the second branch, the defendant must prove that it took all reasonable steps to avoid the particular prohibited event.

As such, the primary defence in the prosecution of OHS regulatory offence is the defence of due diligence. Other defences known in law may also be available to an OHS offence. The basis of the defence is that it would be legally and morally improper to convict a person of an offence when they have taken all reasonable precautions to ensure compliance with the applicable OHS legislation.

In the development of OHS law, there arose an issue with respect to the availability of the due diligence defence. In *R. v. Cancoil Thermal Corp.*[27], it was confirmed that the common law defence of due diligence, as defined in the *Sault Ste. Marie* case, is always available in a regulatory charge under Canadian OHS law. The court dealt with the issue of the specific exclusion of the statutory defence of due diligence in the cases of an offence under the Ontario OHS legislation. The law stated that certain OHS offences were offences of absolute liability, taking away the defence of due diligence. In answering this question, the court ruled as follows:

> ...if section 14(1)(a) [now section 25(1)(a)] were treated as creating an absolute liability offence, it would offend s. 7 of the Canadian *Charter of Rights and Free-*

---

[27]   (1986), 27 C.C.C. (3d) 295 (Ont. C.A.).

doms, the right to life, liberty and security of the person and the right not to be deprived thereof except in accordance with the principles of fundamental justice.[28]

The court went on to comment that since under section 66(1)(a), a violation of section 25(1)(a) of the applicable health and safety legislation may attract a term of imprisonment, the combination of absolute liability and the potential penalty of imprisonment was a violation of section 7 of the *Charter*. The court therefore concluded that to avoid a violation of section 7 of the *Charter*, section 25(1)(a) of the applicable OHS legislation must, at minimum, be treated as creating a strict liability offence.

Proof of the first branch of the due diligence defence is facilitated by meeting the test set down by the Supreme Court of Canada in the *Sault Ste. Marie* case. However, in order to establish the first branch of the due diligence defence, also known as the mistaken fact branch, the following elements must be proven:

1. The accused believed in a mistaken set of facts;
2. If the mistaken set of facts were true, they would render the act or omission innocent;
3. The belief by the accused was deemed reasonable by the court.

The most important decision interpreting the first branch of the due diligence defence in OHS law in the last decade is prosecution in *Ontario v. London Excavators & Trucking Ltd.*[29] In that case, the corporate accused had been hired by a general contractor to perform excavating services for the extension on a new hospital. The equipment operator came into contact with a concrete structure that was not on any of the design drawings. The equipment operators stopped the backhoe, and made an enquiry of a representative of the general contractor. He was advised by the representative of the general contractor that the concrete object was part of a footing of an old nursing station and that it should be removed. However, when the backhoe operator resumed his excavating activity, and dug into the concrete structure, he severed several major power cables from a local hydro utility that were connecting the hospital to the power grid. Although no one was seriously injured, London Excavators & Trucking Company Ltd. was prosecuted by Ontario's Ministry of Labour. At court, the company relied upon the first branch of the due diligence defence. However, the Court of Appeal for Ontario, to

---

[28]   *Ibid.* at 299.
[29]   (1998), 40 O.R. (3d) 32 (C.A.).

which lower court decisions were appealed, held that the defence had not been made out. The Court of Appeal held that before beginning the work, more detailed and objective inquiries should have been made by the excavating contractor. Further, when the unexpected contact was made with the concrete structure, the excavating contractor could not merely rely upon the word of a supervisor of the general contractor, but had to establish more reliably that there was no safety hazard. The court held that it was not objectively reasonable for the backhoe operator, on behalf of London Excavators & Trucking Ltd., to simply rely on the statements made by the supervisor of the general contractor once an unexpected concrete obstacle was encountered. A lower court conviction was upheld by the Court of Appeal.

Proof of the second branch of the due diligence defence is facilitated by compliance with the internal responsibility system and the development of an effective OHS management system. The internal responsibility system of shared duties and responsibilities of various workplace stakeholders is manifest in an effective OHS management system. Workers, as well as employers and supervisors, have duties to comply with the applicable OHS legislation and its regulations. Therefore, many courts have held that an employer is not liable when the circumstances of an accident were attributable to the inadvertence, mistake or negligence of an employee.

One helpful example of a successful due diligence defence was that found in the prosecution of *R. v. Kenaidan Contracting Ltd.*[30] The court had held that the prosecution had proven a contravention of the applicable provisions of the OHS statute in Ontario with which the accused had been charged. However, the court went on to assess whether or not the company had made out the second branch of the due diligence defence. Based on the following factors, the court concluded that the company had established the second branch of the due diligence defence, often referred to as the reasonable precautions branch:

1. There had been a pre-construction meeting in which workplace health and safety was considered as part of the overall planning of the project;

2. A supervisor employed by the contractor was on site at all times to deal with subcontractor issues including health and safety issues;

---

[30] Unreported decision of Justice of the Peace Kitchen, January 12, 1995.

3. An external health and safety consultant had been retained to inspect the project, from time to time, to ensure compliance with the established health and safety programs;
4. There were regular health and safety meetings held at a construction project with all workers;
5. Senior representatives from the contractor attended a weekly project meeting, at which health and safety on the project was an issue;
6. The contractor's supervisor and health and safety consultant had clear authority to stop work if any worker was put in danger;
7. The contractor was never made aware of the concerns relating to the issue of a subcontractor that related to the full-arrest charges against the contractor;
8. The court determined that the health and safety program and the presence of a full-time supervisor were adequate to meet the requirements of the employer under the applicable OHS statute and that the employer had done all that could reasonably be expected of it in the circumstances.

The two branches of the due diligence defence, according to *R. v. Sault Ste. Marie*, clearly placed an onus of proof on the accused to prove the defence. That this was a departure from the long standing Anglo-Canadian legal presumption of innocence was significant. The Supreme Court of Canada indicated in *Sault Ste. Marie* that the accused had the onus of proof to demonstrate that it made out one of the defences in the two branches of the due diligence defence. The standard required by the accused was a civil standard of proof that is proof on a balance of probabilities.

The validity of the reverse onus on the accused to prove the defence of due diligence has been the subject of some legal controversy and challenge. The *Charter*, it was argued, should not permit an accused to have to prove its innocence. The reverse onus of the due diligence defence had exactly that effect. In *R. v. Wholesale Travel Group Inc*,[31] after the Court of Appeal for Ontario rendered the offending provision in the *Competition Act* of no force or effect, the same issue came up in a prosecution under the Ontario *Occupational Health and Safety Act* in *Ellis-Don Ltd. v. Ontario (Labour Relations Board)*[32] In the *Ellis-Don* case, the question of whether the reverse onus on an accused charged

[31] (1989), 63 D.L.R. (4th) 325, 70 O.R. (2d) 545 (C.A.), varied [1991] 3 S.C.R. 154.
[32] 6 Admin. L.R. (3d) 187 (C.A.), Aff'd [2001] 1 S.C.R. 221.

under the OHSA was in violation of subsection 11(d) of the *Charter* was directly raised for the first time in health and safety prosecution. The *Ellis-Don* case became a test case where three cases were joined together by special order of Justice Dubin, then Chief Justice of Ontario. The Court of Appeal gave a split decision, two to one, in favour of the view that the reverse onus on the accused was a violation of the presumption of innocence found in subsection 11(d) of the *Charter*, and that it could not be saved or justified by section 1 of the *Charter*.

The Crown appealed both *Wholesale Travel Group* and the *Ellis-Don* case to the Supreme Court of Canada. The former was heard and decided first by the Supreme Court of Canada. In *Wholesale Travel Group*, the Court reversed the Court of Appeal for Ontario decision and held that the statutory and common law reverse onus on accused to prove, on a balance of probabilities, the defence of due diligence did not offend the *Charter*. The Court was divided on the question of whether there was a contravention of the *Charter* and whether, if this was the case, the infringement was saved and justified by section 1 of the *Charter*. However, the arguments to ultimately justify the reverse onus on the accused in a strict liability offence were generally accepted by the majority of the Court to include the nature of the offence and the policy reasons behind the prosecution of employers for false advertising, polluting and health and safety offences. Placing a reverse onus on an accused in a regulatory, strict liability offence was held not to contravene the *Charter*. Justice Cory identified a so-called licensing justification and a vulnerability justification for the denial of the full presumption of innocence for an accused charged with a strict liability offence, such as an OHS offence.

With the Supreme Court of Canada in the *Wholesale Travel Group* case permitting a reverse onus, the subsequent result in *Ellis-Don* came as no surprise. In *Ellis-Don*, the Supreme Court of Canada released a very brief decision allowing the Crown's appeal. In so doing, the Court adopted its own reasoning from *Wholesale Travel Group*, and rejected the argument that an accused's right under subsection 11(d) of the *Charter* is paramount over the concerns of the OHS regulator in enforcing strict liability offences. Therefore, an accused in an OHS offence is required to prove one or both of the branches of the due diligence defence.

Essentially, when an individual accused or a corporate accused are charged with an OHS offence, they are presumed innocent until such time as the Crown has proven a *prima facie* case, beyond a reasonable doubt. Once the Crown has discharged its burden to prove the prohibited act or omission beyond a reasonable doubt, then the burden of proof shifts to the accused to prove the defence of due diligence. Although the standard of proof on the accused is a civil standard, a balance of probabilities, rather than a criminal standard, the placing of any burden of proof on the accused in the courts of its trial may be argued to be a compromise of the accused's *Charter* right of the presumption of innocence. However, the Supreme Court of Canada has held that is an acceptable requirement for a strict liability offence and does not infringe the *Charter*.

Enforcement of OHS laws in Canada by way of prosecution may, as in a criminal prosecution, result in a conviction and sentencing hearing. Once an accused is convicted of an OHS offence, then the accused is sentenced after submissions from both the crown prosecutor and the defence lawyer. The following chart sets out the current penalties, including fines and jail terms, for an accused convicted of a Canadian OHS offence.

Jurisdiction	Maximum Fine for Organizational Accused[33]	Maximum Fine for Individual Accused
Federal[34]	$1,000,000 or two years in jail, or both.	$1,000,000 or two years in jail, or both.
British Columbia[35]	**First Conviction** $547,229.80 and an additional $27,361.50 for each day the offence continues or 6 months in prison, or both. **Subsequent Conviction** $1,094,459.59 and an additional $54,722.98 for each day the offence continues or 12 months in prison, or both.	**First Conviction** $547,229.80 and an additional $27,361.50 for each day the offence continues or 6 months in prison, or both. **Subsequent Conviction** $1,094,459.59 and an additional $54,722.98 for each day the offence continues or 12 months in prison, or both.

---

[33] *Interpretation Act* R.S.C. 1985, c. I-21, s. 35(1)(11).
[34] R.S.C. 1985, c. L.2, s. 148.
[35] R.S.B.C. 1996, c. 492.

Jurisdiction	Maximum Fine for Organizational Accused	Maximum Fine for Individual Accused
Alberta[36]	**First Conviction** $500,000 and an additional $30,000 for each day the offence continues or 6 months in prison, or both. **Subsequent Conviction** $1,000,000, and an additional $60,000 for each day the offence continues or 12 months in prison, or both.	**First Conviction** $500,000 and an additional $30,000 for each day the offence continues or 6 months in prison, or both. **Subsequent Conviction** $1,000,000, and an additional $60,000 for each day the offence continues or 12 months in prison, or both.
Saskatchewan[37]	$300,000 or two years in prison, or both.	$300,000 or two years in prison, or both.
Manitoba[38]	**First Conviction** $150,000 and an additional $25,000 for each day the offence continues or 6 months in prison, or both. **Subsequent Conviction** $300,000 and an additional $50,000 for each day the offence continues or 6 months in prison, or both.	**First Conviction** $150,000 and an additional $25,000 for each day the offence continues or 6 months in prison, or both. **Subsequent Conviction** $300,000 and an additional $50,000 for each day the offence continues or 6 months in prison, or both.

---

[36] *Occupational Health and Safety Act*, R.S.A. 2000, c. O-2, s. 41.
[37] R.S.S. 1993, c. O-1.1, s. 58.
[38] C.C.S.M. c. W210, s. 55.

Jurisdiction	Maximum Fine for Organizational Accused	Maximum Fine for Individual Accused
Ontario[39]	$500,000	$25,000 or 12 months in prison, or both
Quebec[40]	**First Conviction** $20,000 **Subsequent Conviction** $50,000	**First Conviction** $1,000 **Subsequent Conviction** $2,000
New Brunswick[41]	$50,000 or 6 months in jail, or both	$50,000 or 6 months in jail, or both
Nova Scotia[42]	$250,000 or 2 years in prison, or both.	$250,000 or 2 years in prison, or both.
Prince Edward Island[43]	$50,000 or 1 month in prison, or both	$50,000 or 1 month in prison, or both
Newfoundland[44]	$250,000 or 12 months in prison, or both	$250,000 or 12 months in prison, or both
Yukon Territories[45]	**First Conviction** $150,000 or 12 months in prison, or both.	**First Conviction** $150,000 or 12 months in prison, or both.

---

[39] R.S.O. 1990, c. O.1, s. 66.
[40] R.S.Q. 1990, c. O.1.
[41] S.N.B. 1983, c. O.0.2, s. 47(1).
[42] S.N.S. 1996, c. 7, s. 74(1).
[43] R.S.P.E.I. 1988, c. O-1, s. 31(1).
[44] R.S.N.L. 1990, c. O-3, s. 67(2)(3).
[45] *Occupational Health and Safety Act*, R.S.Y. 2002, c. 159, s. 44.

Jurisdiction	Maximum Fine for Organizational Accused	Maximum Fine for Individual Accused
	Subsequent Conviction $250,000 or 24 months in prison, or both.	Subsequent Conviction $250,000 or 24 months in prison, or both.
Northwest Territories[46]	$500,000 or 1 year in prison, or both.	$500,000 or 1 year in prison, or both.
Nunavut Territory[47]	$500,000 or 1 year in prison, or both.	$500,000 or 1 year in prison, or both.

OHS DUE DILIGENCE AND BILL C-45 "REASONABLE STEPS"

There is a similarity between Canadian OHS law, OHS general duty clauses, and the defence of due diligence on the one hand and the new legal duty under Bill C-45 on the other. In the Bill C-45 amendment to the *Criminal Code* in section 217.1, there is a duty to take reasonable steps to protect the safety of workers and the public. This duty is similar to the general duty clauses found in Canadian OHS law and the language used by the courts in the legal defence of due diligence defence. There is no reverse onus in a prosecution of the new crime of OHS criminal negligence. The question remains, however, regarding what constitutes sufficient proof of reasonable steps in the new health and safety crime.

Due diligence is a primary defence available to an accused against charges under any OHS statute. The second branch requires the accused to prove that it took reasonable steps or precautions to ensure compliance with OHS statutory and regulatory standards. In a criminal prosecution of the new offence of OHS criminal negligence, failure of the accused to take reasonable steps is part of what a prosecutor must prove to the crime.

---

[46] R.S.N.W.T. 1988, c. S-1, s. 22(2) and (4), fine is $50,000 if committed by employee instead of employer.

[47] *Ibid.*

There is every reason to suggest that if an employer complies with the new duty under section 217.1 of the *Criminal Code,* an accused must be compliant with the applicable OHS laws. However, reasonable steps will be difficult to determine consistently across the country since OHS statutes vary jurisdiction by jurisdiction. In most provinces, for example, the OHS legislation contains legal requirements for health and safety committees. In other provinces, specifically Quebec and Alberta, committees are not required unless ordered by regulation or the regulator. These OHS legal requirements may serve as a yardstick in an OHS criminal negligence prosecution, to measure whether or not reasonable steps had been taken. However, since there is no national, consistent OHS statute, this makes the interpretation of reasonable steps more problematic. In other words, it is not clear exactly what will amount to reasonable steps under section 217.1 of the *Criminal Code.* What is clear is that there will have to be some development of the case law to establish what will constitute reasonable steps to prevent bodily harm to workers and the public.

Determining what reasonable steps an employer or manager was required to take in the circumstances will be a fact-specific determination. The mere failure to take reasonable steps required under an OHS regulatory statute, or failure to exercise due diligence in an OHS prosecution, will not necessarily result in a conviction for OHS criminal negligence. This is due to the different onus of proof and fault element of a criminal offence from that of an OHS offence. However, it is reasonable to anticipate that the courts will seek guidance from the extensive due diligence jurisprudence arising from OHS prosecutions that have generally imposed a high standard of care on regulated parties such as employers, constructors, prime contractors, supervisors, officers and directors.

What, then, does the addition of *Criminal Code* OHS duty mean for already regulated organizations and individuals in Canada? Aside from the obvious addition of new criminal liability for organizations, one of the most important differences between criminal and OHS prosecutions is that a criminal standard must be breached by the individual, corporation, or organization. The criminal standard includes proof of wanton or reckless disregard for the lives or safety of others. The criminal standard also requires all elements of the offence to be proven by the prosecutor beyond a reasonable doubt.

Criminal negligence standards have been well developed and were reviewed in chapter 3 of this book. They will continue to evolve with the development of the Bill C-45 jurisprudence. Criminal negligence occurs when an act or omission of an accused party shows wanton or reckless disregard for the lives or safety of others in a situation where the accused party is under a legal duty to act. In respect of organizational liability, the Crown must prove that the conduct of the senior officer of an organization represented a marked departure from the standard of care that could be expected of a reasonably prudent person in the circumstances to prevent a representative from being a party to the offence. This standard is distinct from the regulatory standard of failure to exercise all reasonable care or due diligence. Just how distinct it will be from regulatory due diligence concepts will be dependent on the case law that develops.

CHAPTER 6

CRIMINAL CODE ENFORCEMENT PROCEEDINGS

INTRODUCTION TO THE CRIMINAL JUDICIAL SYSTEM

The amendments to the *Criminal Code* under Bill C-45 increases the scope and application of the criminal liability of organizations and individuals in Canada. Accordingly, a large number of persons may be exposed to criminal investigations and charges that have had no previous experience with the criminal justice system. Faced with this prospect, it is imperative to have a basic understanding of the Canadian criminal justice process as well as the fundamental rights the Canadian Constitution affords to each accused. This chapter gives a brief description of the salient authority, stages of the process and legal rights that an accused has during the criminal justice process, from the time the police commence a criminal investigation to the time an accused is either acquitted or convicted and remanded for sentencing.[1]

The new offence of OHS criminal negligence is a criminal offence and will be prosecuted under the Canadian criminal justice system. Under the Canadian Constitution, only the federal parliament may enact laws concerning criminal law and procedure.[2] Most criminal law in Canada is contained in the *Criminal Code* of Canada and under the *Controlled Drugs and Substances Act*.[3] Not all laws enacted by the

---

[1]   Note that the following discussion is intended to help persons criminally charged under the provisions of the *Criminal Code* that will be amended or added through the enactment of Bill C-45. Also, the areas of procedure described below do not necessarily form part of every case. They provide a general overview of common stages. Every criminal proceeding has a life and history of its own and will be influenced by the particular facts of each case and the procedural decisions and options available.

[2]   *Constitution Act, 1867,* (U.K.), 30 & 31 Vict., c. 3, s. 91 (27).

[3]   S.C. 1996, c. 19.

federal government, however, are enacted pursuant to its criminal law power. For example, the *Canada Labour Code*, Part II, regulating occupational health and safety in federally regulated workplaces, is passed pursuant to the federal government's authority to regulate federal works and undertakings, not criminal law. Most OHS laws in Canada are passed under the authority of provincial jurisdiction.

Canadian provinces, unlike countries such as Australia and the United States, are not permitted to make laws that are classified criminal law. Criminal offences are established when the federal government passes a law that has moral blameworthiness and a fault element as part of the offence. It need not be passed as part of the *Criminal Code* to be a criminal law. For example, if a municipal by-law has as its primary purpose the punishment of behaviour as a criminal act then it would be held to be unconstitutional.

The attorney general of a province prosecutes *Criminal Code* offences. The federal attorney general prosecutes drug offences. The same division of powers distinction is made with respect to policing. Provinces have authority to establish police forces to enforce the *Criminal Code*. The federal government has authority to establish a police force to investigate drug offences and other federal offences. However, many provinces contract for the services of the national police force, the Royal Canadian Mounted Police. Only Ontario and Quebec have provincial police forces. The *Criminal Code* gives special powers to peace officers, which is another term for police officers in Canada, to enforce criminal law. The *Charter*, previously discussed, is part of the fundamental law of Canada and attorney generals and peace officers are bound to comply with the *Charter* when enforcing the *Criminal Code* or OHS offences.

The majority of criminal cases are resolved in provincial courts, usually with a plea of guilty. Provincial court judges are appointed by the provinces and they do not sit with a jury. Provincial court judges may hear serious indictable offences, provided that the accused decides or chooses to be tried in a provincial court without a jury. In such circumstances, there is no preliminary inquiry. Federal parliament has exclusive authority to appoint superior court judges, who can sit with a jury. Murder charges must be tried by a superior court judge, usually with a jury. Superior court judges may also try most other indictable offences, provided that the accused elects or chooses to be tried in this higher level of trial court. The accused can elect to be tried in the superior court, with a preliminary inquiry, and with or without a jury.

The *Criminal Code* applies across Canada and sets the standard of what constitutes criminal behaviour in every part of the country.[4] There is no comparable OHS statute of national application. In this respect, the *Criminal Code* is consistent in setting standards of criminal behaviour throughout Canada. No person may be convicted or discharged of a *Criminal Code* offence committed outside of Canada.[5] The *Criminal Code* provides for some exceptions for offences on an aircraft, ship, fixed platforms and spacecraft that are outside of Canada but that have a designated connection with Canada.[6]

THE CLASSIFICATION OF OFFENCES

The *Criminal Code*, Canada's primary criminal law statute, defines an act or omission to be a criminal offence and how the offence is classified. All crimes in the *Criminal Code* are classified as either indictable offences or offences punishable upon summary conviction. The former represent the more serious offences with more severe penalties. Additionally, some crimes have a dual or hybrid character of being both indictable and summary conviction offences. This means that in some cases the Crown has the right to select how the offence will be classified and prosecuted.[7] Neither the court nor the defence has the right to direct the Crown how it must proceed. However, if the Crown fails to elect how it will proceed, the offence is deemed to be an indictable offence for the purposes of any interim procedures such as bail hearings or other motions.[8]

Summary Conviction Offences[9]

Summary conviction offences are offences that are less serious on the spectrum of offences under the *Criminal Code*. Most sections of the *Criminal Code* enacting offences punishable on summary conviction

---

[4] Although the *Criminal Code* is the pre-eminent source for determining offences in Canada, there are also a number of other federal offences declared in other federal statutes and regulations, as well as provincial offences such as driving and liquor offences declared in provincial statutes and regulations.

[5] R.S.C. 1985, c. C-46, s. 6(2).

[6] *Ibid.*, s. 7.

[7] "Crown" is used interchangeably with "prosecution" throughout this text.

[8] *Interpretation Act*, R.S.C. 1985, c. I-21, s. 34(1)(a).

[9] Proceedings for summary conviction offences are governed by Part XXVII of the *Criminal Code*.

often contain no specific penalty provisions. If this is the case, except where otherwise provided by law, every one who is convicted of an offence punishable on summary conviction is liable to a fine of not more than $100,000 or to imprisonment for six months or both.[10] This penalty provision was specifically amended by Bill C-45 by increasing that monetary penalty from $25,000 to $100,000 for a summary conviction offence. Nevertheless, if provided for in the applicable section of the *Criminal Code*, some offences prosecuted by way of summary conviction will have a maximum penalty of imprisonment of eighteen months.[11]

Indictable Offences

As mentioned above, indictable offences are the most serious offences and accordingly, their penalties are the most severe. They are punishable by imprisonment for two, five, ten, fourteen years or life imprisonment depending on the offence. An accused who is charged and convicted of the new offence of OHS criminal negligence causing death may receive a maximum penalty of life imprisonment. The *Criminal Code* provides for three categories of indictable offences:

(i)   *Absolute Jurisdiction or section 553 Offences*: This class of indictable offences fall within the absolute jurisdiction of a provincial court judge. The accused must be tried in the provincial court of the province or territory.[12]

(ii)   *Section 469 Offences*: This class of indictable offences fall within the exclusive jurisdiction of the superior court of the province or territory.[13]

(iii) *Section 554 Offences*: All other indictable offences which do not fall within section 469 or section 553 allow an accused an election as to the mode of trial (discussed below). These include fraud over $5,000, criminal negligence causing death, manslaughter, etc.

Whether a criminal prosecution is by way of summary conviction or indictment, the criminal investigation process is usually the same. The

---

[10]   *Ibid.*, s. 787(1).
[11]   *Ibid.*, s. 264.1(a) (uttering a threat to cause death or bodily harm).
[12]   These are less serious offences; for example, theft under $5,000.
[13]   These offences are serious; for example, murder.

criminal process usually commences before an arrest is made as investigations must be conducted by the police. Police usually do this by way of investigation, search and seizure, surveillance and witness interviews. Correspondingly, potential accused persons or organizations have various pre-charge and post-charge rights during the criminal investigation process. Some of these rights of accused are reviewed below.

THE RIGHT TO BE FREE FROM UNREASONABLE SEARCH AND SEIZURE

It is the fundamental right of every Canadian individual and organization to be secure against unreasonable and arbitrary searches by the police and to not have their property seized for use as evidence as a result of such searches. This right, firmly rooted in our common law tradition, is now codified as a constitution right in section 8 of the *Charter*. The Supreme Court of Canada has interpreted the purpose of section 8 of the *Charter* to be the protection of people's reasonable expectations of privacy.[14] The law generally obliges no person to engage in self incrimination. This includes warrantless searches in one's home or place of business, body searches for possible production of DNA evidence,[15] the acquisition of someone's or a business' hydro information or bank records. The right to be free from unreasonable search and seizure, under section 8 of the *Charter*, in taking forensic DNA evidence, is protected by section 487.05 of the *Criminal Code*. That provision permits a provincial court judge to issue a DNA warrant for taking bodily samples for forensic DNA analysis. A provincial court judge may issue a DNA warrant where there are reasonable grounds to believe that a designated offence has been committed and that a bodily substance has been found or obtained at the scene, on or in the victim, or on anything worn or carried by the victim, or on or within the body of any person or anything or at any place associated with the commission of the designated offence. The provincial court judge, in considering a request for a DNA warrant, must consider the nature of the designated offence and the circumstances of its commission and where there is a peace officer or other person with the necessary training who is able to take the bodily sample from the suspect.

---

[14] See *Hunter v. Southam Inc.*, [1984] 2 S.C.R. 145.
[15] See *R. v. Stillman*, [1997] 1 S.C.R. 607.

Every invasion of private property, however slight, is considered to be a form of trespass. Generally speaking, no person has the right to enter private property except by the owner's or occupier's consent, or strictly in accordance with particular lawful authorization. Unless the police are in possession of a valid search warrant or are acting in accordance with some other specific legal authority,[16] they have no right to enter private premises and remain there against the owner's or occupier's wishes. When police obtain a search warrant, it must meet certain substantive requirements. The search warrant must state an offence with sufficient precision to apprise anyone concerned with the nature of the offence for which evidence is being sought; it must describe the items to be seized with enough specificity to permit the officers responsible to execute the search warrant to identify such objects and to link them to the offence described in the information and the search warrant; it must describe the location with sufficient accuracy to enable one from the mere reading of it to know of what premises it authorizes the search.[17]

Evidence acquired as a result of an illegal search and seizure may be subject to an application for exclusion from being used as evidence at trial if it may affect the fairness of the trial.[18] However, if the prosecution can show that illegally obtained evidence was inevitably to be found, the evidence may be admissible. The court may be called upon to determine, under subsection 24(2) of the *Charter*, whether the admission of the illegally obtained evidence brings the administration of justice into disrepute. Subsection 24(2) requires that evidence shall be excluded if, having regard to all of the circumstances, the admission of the evidence would bring the administration of justice into disrepute. In addition to remedies under subsection 24(2) of the *Charter*, evidence may also be excluded under subsection 24(1) of the *Charter*. For example, if the admission of evidence undermines an accused's right to a fair trial, pursuant to subsection 11(d) of the *Charter*, then an order excluding the

---

[16] In *R. v. Collins* (1987), 33 C.C.C. (3d) 1 (S.C.C.), the Court established a five-step process to determine the legality of a search and seizure: 1) can the police (or State actor) conduct be characterized as a search and seizure? 2) Did the actions of the police (or State actor) intrude upon the accused's reasonable expectation of privacy? 3)Were the actions of the police authorized by law, either statutory or the common law? 4) Was the law that the police were relying on to pursue the search and seizure reasonable? 5) Was the search or seizure carried out in a reasonable fashion? The accused has the burden of persuading the court that his or her *Charter* rights against unreasonable searches has been infringed or denied.

[17] *R. v. Gibson*, [1919] 30 C.C.C. 308 (Alta. T.D.).

[18] *Canadian Charter of Rights and Freedoms, Constitution Act, 1982*, being Schedule B to the *Canada Act*, 1982 (U.K.), c. 11, s. 24(1).

evidence may be obtained by the trial judge. Evidence may be rendered unfair at trial, the way it was taken may render it unreliable or its potential for misleading the trier of fact may outweigh any such minimal probative value it may possess. Further, if the police have acted in such an abusive fashion that the court concludes that the admission of the evidence would irremediably taint the fairness of the trial itself, it may also be excluded under the authority of subsection 11(1) of the *Charter*.[19]

THE RIGHT TO BE FREE FROM ARBITRARY DETENTION[20]

A detention occurs when a person has been taken into police custody or believes that he or she does not have the option to leave when being detained or questioned. Many people do not know exactly what the authority of the police is and will tend to err on the side of co-operation rather than caution when dealing with them. When confronted by police, even if feeling uncomfortable and pressured, individuals will often decide not to leave or defer questioning. The law clearly states that any person has the right not to be arbitrarily detained by the police. If a person was unlawfully detained by the police, without legal representation, it is possible that any evidence gathered was obtained contrary to section 10(b) of the *Charter*. The corresponding right of a detained person to retain and instruct a lawyer without delay and to be informed of that right is an important pre-charge right of every individual under the *Charter*.

Organizations, including corporations, cannot be physically detained. Nonetheless, if any of its managers or representatives are put in such a situation, they will be allowed, on behalf of the organization, to retain and instruct counsel. This will be of increasing importance with potential risk of criminal prosecution of OHS criminal negligence.

INITIATION OF CRIMINAL PROCEEDINGS

Criminal proceedings, following an investigation, are usually commenced when an accused person or organization is formally charged. This may be done by the prosecution in one of two ways: 1) on the swearing of an information or 2) the issuance of an indictment. It is the Crown prosecutor who will decide what charges should be laid. If criminal charges proceed, the accused should retain a lawyer or, if a

---

[19] *R. v. Harrier* (1995), 101 C.C.C. (3d) 193 (S.C.C.)
[20] *Ibid.*, s. 9.

lawyer cannot be afforded, apply for legal aid to have a lawyer appointed by the provincial legal aid system.

An Information is generally prepared for summary conviction offences or offences of a less serious nature that are to be tried in the provincial court. The Information records the offence and is used to record the progress of the case. It contains information such as whether the accused was released on bail, what sort of release the accused was given, how the accused pleaded, the disposition of the case if it was tried in the provincial court, the outcome of the preliminary inquiry if the accused elected trial in the superior court, and the time and date of every appearance in court.

An indictment is a written accusation generally signed by the prosecutor in the name of the Queen alleging that an accused has committed an indictable offence. It is the formal legal document, like the Information, containing the records and progress of the case, and used throughout the criminal proceedings which will be heard in the superior court of criminal jurisdiction.

ARRESTING AN ACCUSED

Individuals are often arrested when they are charged with a criminal offence.[21] Section 494 of the *Criminal Code* sets out the arrest powers that anyone has without a judicial warrant and section 495 sets out the wider arrest powers of a police officer without a warrant. The general rule, as provided for in section 495(2), is that a peace officer shall not arrest a person without warrant for less serious offences under the *Criminal Code* where, in establishing the identity and securing the attendance of the person in court, securing evidence of the offence and preventing the repetition of the offence or the commission of another offence, the public interest that justice will be achieved can be satisfied by issuing an appearance notice.[22]

A judicial warrant for an arrest may be obtained under section 507(4) of the *Criminal Code*. In such instance, if a justice of the peace[23]

---

[21] For further information, see M.L. Friedland and K. Roach, *Criminal Law and Procedure: Cases and Materials,* 7th ed. (Toronto, Emond Montgomery Publications Limited, 1994).

[22] This is essentially a promise to appear by the accused. Should the accused not appear, he may be arrested and this will affect his or her chances of receiving bail in the future.

[23] A Justice of the Peace is the person who presides over the case, much like a judge but with a much narrower scope of authority.

considers that a case has been made out on oath for compelling an accused to answer the charge of the offence, then he or she will issue a summons unless the evidence presented discloses reasonable grounds to believe that it is necessary in the public interest to issue an arrest warrant.

The exercise of arrest powers triggers a number of ancillary powers. The Supreme Court of Canada has held that in the exercise of their arrest powers for indictable offences, police officers may enter premises without consent if there are reasonable and probable grounds to believe that the person sought is within the premises and if, before entry, they have provided proper announcement for their presence and purpose.[24] This common law power has been restricted under the *Charter*. Also, when exercising arrest powers, police officers may make warrantless searches incidental to the arrest.[25] The Supreme Court of Canada in the decision of *R. v. Feeney*[26] held that the common law right to make a warrantless arrest on private premises must now be adjusted to comply with *Charter* values. In general, an individual's privacy interest in the dwelling-house outweighs the interest of the police, and warrantless arrests in dwelling-houses are prohibited under the *Charter*. Under section 529.1 of the *Criminal Code* a justice or judge is allowed to issue a warrant to authorize a peace officer to enter a dwelling-house for the purpose of effecting the arrest or apprehension of a suspect. This judicial authorization is dependent upon the justice or judge being satisfied by information on oath in writing that there are reasonable grounds to believe that the person is or will be present and that an arrest warrant is in force anywhere in Canada or that grounds for a warrantless search exist. This provision, as well as the companion provisions in sections 529.2 through to 529.5 inclusive, were enacted in response to the Supreme Court of Canada decision in *R. v. Feeney.*

If a person is arrested, summoned or given an appearance notice for an indictable offence, he or she may be compelled to attend at a specified place and time for photographing and fingerprinting under the *Identification of Criminals Act.*[27]

In general, after an arrest is made, a peace officer is under an obligation to release the arrestee from custody as soon as practicable and issue a summons or appearance notice[28] unless the peace officer believes

---

[24] *Eccles v. Bourque* (1974), 19 C.C.C. (2d) 129; *R. v. Landry* (1986), 25 C.C.C. (3d) 1.
[25] M.L. Friedland and K. Roach, *supra* note 21 at 114.
[26] (1997), 115 C.C.C. (3d) 129 (S.C.C.)
[27] R.S.C. 1985, c.I.1.
[28] *Criminal Code*, s. 497.

on reasonable grounds that it is necessary in the public interest to detain the accused in custody having regard to the need to establish the identity of the arrestee, secure or preserve evidence relating to the offence, prevent the continuation of an offence, or to stop the arrestee from failing to appear in court.

It is trite to say that organizations, including corporations, cannot be physically detained or arrested. Rather, they must be served with a summons or notice that they have been charged with a criminal offence under the *Criminal Code* or any other act.[29] If no other method of service is provided, service on an organization may be effected by delivery to the manager, secretary or other executive officer of the organization or to any one of its branches.[30]

RIGHT TO COUNSEL

As previously referenced above, the criminal investigative activities of the police are restricted by various rights under the common law and under the *Charter*, including the right to counsel under section 10(b). Section 10 states that:

> every one has the right on arrest or detention:
>
> > (*a*) to be informed promptly of the reasons therefore;
> >
> > (*b*) to retain and instruct counsel without delay and to be informed of that right.

The right to legal counsel is a fundamental right of every suspect or accused. The right to legal counsel has been explained by the Supreme Court of Canada as follows:

> [W]hen an individual is detained by state authorities, he or she is put in a position of disadvantage relative to the state. Not only has this person suffered a deprivation of liberty, but also this person may be at risk of incriminating him- or herself. Accordingly, a person who is "detained" within the meaning of s. 10 of the *Charter* is in immediate need of legal advice in order to protect his or her right against self-incrimination and to assist him or her in regaining his or her liberty.[31]

Once an accused has expressed his or her desire to retain a lawyer, the police cannot question him or her until he or she has had a meaningful opportunity to speak with a lawyer. The same right is afforded to organizations charged with a crime. Through its directors or officers, the

---

[29] *Ibid*, s. 703.2.
[30] *Ibid*.
[31] *R. v. Bartle*, [1994] 3 S.C.R. 173 at para. 16.

organization can consult with a lawyer to ascertain its rights. Nonetheless, once an accused has received advice from his or her lawyer, the police can resume their questioning. Therefore, an accused should remember that at all times he or she is under no obligation to say anything to the police and that by not saying anything, he or she will not be adversely prejudiced later.[32]

It is a *Charter* violation if the police tells an accused that he has a right to a lawyer but then proceeds to tell the accused that he or she does not really need a lawyer.[33] By doing so, a police officer has been held by the courts to be depriving the accused of the right to make a meaningful decision because the police officer is providing advice. Further, if an accused expresses concern about his or her ability to afford a lawyer, the police must inform him or her of the availability of legal aid and/or duty counsel.[34]

JUDICIAL INTERIM RELEASE – A.K.A. "BAIL"

The hearing for judicial interim release, also known as bail, is a very important part of the criminal process. The *Criminal Code* permits the police to initially determine an accused's pre-trial release status. There are certain instances where it is necessary for an accused to be held in custody, either because the police have decided not to release the accused pursuant to sections 495 to 499 of the *Criminal Code* or because the police are unable to release the accused, as in a case of murder. In such cases, the accused must be brought before the courts.

Generally, there are two types of accused for the purposes of the judicial interim release provisions: those charged with an offence listed in section 469 of the *Criminal Code*, most frequently murder, and those charged with any other criminal offence. For those charged with offences not listed in section 469, an automatic right to a bail hearing results and they must be brought before a court[35] without unreasonable delay and in

---

[32] *Charter*, s. 7.

[33] *R. v. Burlingham* (1995), 124 D.L.R. (4th) 7 (S.C.C.).

[34] Duty Counsel is a government lawyer who is available before court to advise accused persons. It is provided to ensure that an accused person will have the benefit of legal advice and assistance when appearing before the Court unrepresented. All accused persons are entitled to assistance by Duty Counsel regardless of their financial means. The financial eligibility guidelines governing the issuance of Legal Aid certificates does not apply to the Duty Counsel program.

[35] Note that the type of offence will determine which judicial stream the accused will enter for the purposes of bail (appearance before a provincial court or superior court).

any event within 24 hours if a justice of the peace is available.[36] At a bail hearing, the justice of the peace is obliged to order the accused released on an undertaking without conditions with respect to the offence unless the prosecutor, having been given a reasonable opportunity to do so, shows cause why the detention of the accused in custody is justified or why a more onerous form of release ought to be ordered.[37] The criteria for release/detention are dictated by subsection 515(10) of the *Criminal Code*. They include: 1) where the detention is necessary to ensure his or her attendance in court, commonly referred to as the primary ground; 2) where the detention is necessary for the protection or safety of the public, commonly referred to as the secondary ground; or 3) where the detention is necessary in order to maintain confidence in the administration of justice, commonly referred to as the tertiary ground.

Accused persons charged with section 469 offences are not entitled to an automatic release hearing before a justice. On the contrary, unlike others, they must make an application for release to a judge of the superior court and bear the onus of showing cause as to why they ought to be released.[38] If they are successful in doing so, the justice must order their release and include in the record a statement of reasons for making the order.[39] The justice may also include any conditions available on the release.[40]

---

[36] *Criminal Code*, s. 503.

[37] *Ibid.*, s. 515(1)(2).

[38] This reverse onus situation is also true for accused persons charged with: offences under *Criminal Code* ss. 467.11, 467.12 or 467.13, or an offence alleged to have been committed for the benefit or in association with a criminal organization; an offence under any of sections 83.02 to 83.04 and sections 83.18 to 83.23 or otherwise is alleged to be a terrorism offence; an offence under subsections 16(1) or (2), 17(1), 19(1), 20(1) or (22(1) of the *Security of Information Act*; an indictable offence not listed in section 469 where the accused is not ordinarily resident in Canada (as per s. 515(6)(b)); an offence under subsections 145(2) to (5) alleged to have been committed while the accused was at large after being released in respect to another offence; having committed an offence punishable by sections 5(3), 5(4) or 6(3) of the *Controlled Drugs and Substances Act,* S.C. 1996, c.19 (as per subsection 515(6)(d) of the *Criminal Code*).

[39] *Criminal Code*, s. 515(6).

[40] These include: the supervision of the accused by a surety or sureties or, if the prosecution consents, the deposit of cash bail in lieu of finding a surety or sureties (s. 515(2)); the reporting to a designated person, such as a police officer (s. 515(4)(a)); remaining within a specified territorial jurisdiction (s. 515(4)(b)); notifying the Court of any change of address or employment (s. 515(4)(c)); abstaining from communication with named persons such as victims, co-accused, witnesses; depositing one's passport (s. 515(4)(e); or any other condition that the Court sees fit to impose (s. 515(4)(e.1)).

It is possible that an accused be denied bail and remain in pre-trial detention, no matter what offence he or she is charged with, until the end of their trial. Although time in custody is usually given full credit in sentencing, pre-trial incarceration still amounts to a loss of personal liberty. Legal counsel can advise on matters of bail and represent accused in bail applications.

CROWN DISCLOSURE

Crown disclosure is the process of making the prosecution's file available to the accused and his or her lawyer. Section 7 of the *Charter* provides an accused with a right to make full answer and defence in response to criminal charges. Crown disclosure is one of the most important rights in the criminal trial process and as such, the Crown has an obligation to disclose to the accused all relevant information in its possession, whether it be inculpatory or exculpatory.[41] The Supreme Court has stated, "the fruits of the investigation...are not the property of the Crown for use in securing a conviction but the property of the public to be used to ensure that justice is done."[42] In fact, the Crown has a special obligation to treat an accused fairly and provide crown disclosure. Consequently, crown disclosure must be made early in the prosecution process and ongoing disclosure of new evidence and material is a continuing duty of the Crown. Failure by the Crown to fulfill its disclosure obligation may result in a stay of proceedings or other redress, including costs, and will be treated as a very serious breach of professional ethics.[43]

An accused is entitled to disclosure and production of evidence where he or she can show that there is a "reasonable possibility that the information is logically probative."[44] Accordingly, the Crown must disclose all information and evidence that is in its custody, including statements that witnesses have made and any physical evidence seized. Furthermore, it must also produce information and evidence that is in its power, possession or control, such as evidence still in the hands of the police or other agents of the state that are custodians of information[45]. Correspondingly, the Crown has a duty to obtain from the investigators

---

[41] See *R. v. Stinchombe*, [1991] 3 S.C.R. 326, 68 C.C.C. (3d) 1; see also *R. v. Carosella*, [1997] 1 S.C.R. 80.

[42] *Stinchcombe, ibid.*

[43] *R. v. Chaplin*, [1995] 1 S.C.R. 727, 96 C.C.C. (3d) 225 at 233.

[44] *R. v. O'Connor*, [1995] 4 S.C.R. 411, 103 C.C.C. (3d) 1.

[45] For example, another government agency assisting in the prosecution.

all relevant information and then preserve it.[46] If evidence that should have been disclosed be lost, the prosecution must explain what happened to it and if the reason provided is deemed unsatisfactory, an abuse of process or other breach of section 7 of the *Charter* may have occurred.[47]

If the evidence or documentation is not in the Crown's immediate possession or control, but rather in the possession and control of a third party, the accused can apply to the trial court for an order requiring the third party to disclose the relevant materials needed.[48] In a request for third party disclosure the defence counsel must demonstrate two separate requirements. First, that the material requested is relevant to an issue at trial or the competency of a person to testify at trial. Second, the documents, records, or evidence are produced to the trial court for examination and the trial court then weighs the positive and negative effects of ordering a production of documents, evidence, or records by third parties.

Despite this positive duty of disclosure, the Crown can decide to delay or deny production of evidence for legitimate reasons such as relevancy[49], the protection of informers or witnesses, or the completion of an investigation.[50] Also, disclosure may be denied where justified by the law of privilege. Nonetheless, where such evidence or material is not disclosed to the defence, there remains an obligation on the prosecution to disclose the existence of that material. Finally, full crown disclosure permits the accused and defence counsel the opportunity to know the case it has to meet. This promotes open access to the prosecution's case and hopefully facilitates early resolution of the charges.

THE ACCUSED'S ELECTION IN CRIMINAL NEGLIGENCE

The *Criminal Code* sets out the election process for the venue of a criminal trial. After the Crown has elected how it wishes to proceed with a charge, either by indictment or by summary conviction, the accused generally has a choice as to how he or she prefers the trial to proceed. Unless the offence is one listed under either of sections 553 or 469 of the

---

[46] *R. v. O'Grady* (1995), 64 B.C.A.C. 111, [1995] B.C.J. No. 2041 (C.A.).

[47] *R. v. La*, [1997] 2 S.C.R. 680.

[48] See *O'Connor*, 103 C.C.C. (3d) 1; see also *Criminal Code*, s. 278.2.

[49] Information that is totally irrelevant need not be disclosed, but the Crown must err on the side of inclusion, disclosing all information that might be useful to the defence in making a full Inquiry.

[50] Note that this decision not to disclose evidence, though rarely exercised, can be reviewed by a judge.

Criminal Code, he or she may elect to proceed with a judge alone trial in the provincial court, or have a preliminary inquiry in the provincial court followed by a judge alone trial in the superior court of criminal jurisdiction, or a judge and jury trial in the superior court of criminal jurisdiction. If the accused is charged with an offence listed under section 553, he or she will have no choice but to face a trial by judge alone in the provincial court. Conversely, if the accused is charged with an offence listed under section 469, the superior court of criminal jurisdiction has exclusive jurisdiction over the trial and it is to be preceded by a preliminary inquiry in the provincial court. In such circumstance, the accused will nonetheless be able to select whether he or she wishes to be tried by a judge alone or by a judge and jury,[51] unless the offence is punishable by imprisonment for more than five years, in which case the attorney general may require a trial to proceed by judge and jury.[52]

If a provincial court judge declines to record an accused's election, as is his right under section 567,[53] the accused is deemed to have elected to be tried by a court composed of a judge and jury. A similar result occurs if an accused did not elect when his election was put to him under section 536, or if he was ordered to stand trial by a provincial court judge who continued proceedings before him as a preliminary inquiry, pursuant to subsection 555(1) of the *Criminal Code*.

Where an accused elects to be tried by a provincial court judge without having a preliminary inquiry, the provincial court judge shall call upon the accused to plead to the charge, and if the accused does not plead guilty, the provincial court judge shall proceed with the trial or fix a time for the trial. If the election to be tried by a provincial court judge is made before a justice who is not a provincial court judge, the justice must remand the accused to appear and plead to the charge before a provincial court judge.

Where an accused elects to have a preliminary inquiry and be tried by a judge without a jury, or a court composed of a judge and jury, or where the accused is deemed to have elected to be tried by a court composed of a judge and jury, the provincial court judge before whom the election is made will conduct a preliminary inquiry.

---

[51] Note that section 11(f) of the *Charter* confers upon any person charged with an offence the right, save for offences under military law, to be tried before a judge sitting with a jury where the maximum punishment for the offence is imprisonment for five years or a more severe punishment.

[52] *Criminal Code*, s. 568.

[53] A judge would do so to ease the procedure, if he or she wanted to prevent multiple accused persons from electing different modes of trial.

Sometimes, an accused may wish to change his or her election. If the original election was to be a trial in provincial court, the accused may, as of right, re-elect in writing to have a preliminary inquiry and be tried by a judge with or without a jury if the accused does so not later than fourteen days before the first day of trial.[54] If he or she wishes to do so less than fourteen days before the first day of trial, such re-election requires the consent of the prosecutor.[55]

If the accused had originally elected to be a tried in the superior court, with or without a jury, he or she may re-elect to be tried in provincial court at any time before or after the completion of the preliminary inquiry, provided the prosecutor gives written consent.[56] On receipt of such written re-election, the judge presiding at the preliminary inquiry shall proceed with the trial or fix a date for the trial.[57]

Re-election is also permitted if the accused had originally elected a mode of trial other than in provincial court and wishes to re-elect another mode of trial not in provincial court, provided less than fifteen (15) days have elapsed since the completion of the preliminary inquiry. Should this re-election take place more than fifteen days after the completion of the preliminary inquiry, the written consent of the prosecutor is required.

THE PRELIMINARY INQUIRY

In the case of indictable proceedings, the criminal prosecution process may include the holding of a preliminary inquiry. The purpose of a preliminary inquiry is to determine whether there is sufficient evidence to require that an accused person be placed on trial for the offence they are charged with, or any other indictable offence in respect of the same set of events.[58] Sometimes a preliminary inquiry is held pursuant to an election of the accused, as in the case of an offence where neither a provincial court judge nor a superior court of criminal jurisdiction has absolute jurisdiction.[59] Sometimes a preliminary inquiry is mandatory, as in the case of any offences listed under section 469 of the *Criminal Code*. However, in such a case, an accused may dispense with a preliminary

[54] *Criminal Code*, s. 561(1).
[55] *Ibid.*, s. 561(2).
[56] *Ibid.*, s. 561(1).
[57] *Ibid.*, s. 562(2).
[58] *Ibid.*, s. 548, the test of which is set out in *United States of America v. Shephard*, [1977] 2 S.C.R. 1067.
[59] *Ibid.*, ss. 553 and 469.

inquiry, with the Crown's consent, by consenting to an order to stand trial pursuant to subsection 549(1).

When a preliminary inquiry is held, it is required to proceed in accordance with Part XVIII of the *Criminal Code*. The Crown, to obtain a committal for trial, must show sufficient evidence upon which a reasonable jury properly instructed could convict the accused. If the evidence is insufficient to result in a reasonably minded trier of fact to make a finding of guilt, the accused must be discharged. An accused is entitled to present evidence but is under no obligation to do so on a preliminary inquiry. Usually an accused will use the preliminary inquiry to test the Crown's case against the accused.

A decision of the Court of Appeal for Ontario stands for the proposition that the preliminary inquiry "protects an accused from trial where the Crown is unable to produce sufficient evidence to warrant the accused's committal for trial."[60] A preliminary inquiry affords an accused the opportunity to learn about and discover the case against him or her,[61] to commit a witness to details of his testimony that will later be used in cross-examination at trial should the witness' testimony differ, to challenge the strength of the evidence adduced by the Crown, to probe for details and facts which may allow the defence to unearth facts or witnesses who might otherwise not be available. An accused may also obtain information to prepare for *Charter* motions or other motions at trial at the preliminary inquiry. Conversely, the Crown may be afforded the opportunity to discover the theory of the defence, to test the strength of its evidence, and to record the sworn testimony of potentially reluctant witnesses and commit them to details prior to trial.

A preliminary inquiry may be held for an individual or an organization as an accused as amended by Bill C-45. By virtue of sections 538 and 556 of the *Criminal Code*, an organization appears in court by counsel or agent. The justice may, however, proceed with a preliminary inquiry in the absence of a corporate accused if it does not appear pursuant to a summons.

PRE-TRIAL CONFERENCE

A pre-trial conference is a meeting between prosecution and defence counsel and a judge who will not preside at trial. Before a trial commences, it is sometimes necessary that a pre-hearing conference with

---

[60] *R. v. Girimonte*, 121 C.C.C. (3d) 33 at 46.
[61] *R. v. Skogman*, [1984] 2 S.C.R. 93, 13 C.C.C. (3d) 161 at 171.

a judge, usually referred to as a pre-trial, be held. The *Criminal Code* provides that a pre-trial conference is mandatory in every case to be tried by a jury[62] and may be ordered in non-jury cases.[63] An accused person can also meet with the Crown for a resolution meeting to discuss possible pleas and obtain the Crown's position on a resolution without a trial prior to or even during the trial.

The pre-trial conference is a rather informal proceeding generally used to identify and narrow the outstanding issues and to ensure the efficient use of court time. For such a conference to be productive, an accused should have discussed the case with his lawyer to address issues such as the anticipated length of trial, the completeness of crown disclosure, the making of admissions, arguments concerning the admissibility of evidence and applications for *Charter* remedies. Furthermore, many judges conducting pre-trial conferences will express their views on legal issues, the appropriate penalty or the possibility of guilty plea of the accused to a lesser charge. A pre-trial conference may have the benefit of exploring a possible resolution of the criminal charges without a trial.

TRIAL

A trial is the process in the criminal justice system by which the accused is brought to justice to determine if they are guilty of the charge with which they are charged. Once the procedural steps described above have been completed and the accused has made his or her intention clear to plead not guilty, a trial is ready to commence. A trial is based on the adversarial system in the administration of justice in Canadian courts. An accused should, however, be cognizant of the right to a trial within a reasonable time, under subsection 11(b) of the *Charter*. Should this right be violated, a stay of proceedings pursuant to subsection 24(1) of the *Charter* may be warranted.

The right to be tried within a reasonable period of time is one of the most important rights of an accused. Individuals may suffer exquisite agony waiting for a determination of their criminal charges at trial. If judicial resources are not available for a speedy trial, then this prejudices the right of an accused to be fairly treated in the administration of criminal justice. Corporations have a right under subsection 11(b) of the *Charter* to be tried within a reasonable period of time, as well as an individual accused. However, the Supreme Court of Canada has held that

---

[62] *Criminal Code*, s. 625.1(2).
[63] *Ibid.*, s. 625.1(1).

in addition to an unreasonable delay in proceeding to trial, over and above that of an individual accused, "…a corporate accused must…be able to establish that its fair trial interest has been irremediably prejudiced."[64]

Assuming that the subsection 11(b) right to a trial within a reasonable period of time under the *Charter* has not been violated, the beginning of a trial usually begins with the accused being arraigned. This consists of the accused standing before the court, there is a formal reading of the charge or charges to the accused, and a clerk asks the court how the accused wishes to plead, guilty or not guilty. The accused must then plead either guilty, or not guilty. An accused may also plead not guilty as charged but guilty to a lesser and included offence or other criminal offence. In the case of an individual, the answer to the arraignment should be done personally by the accused so that there can be no misunderstanding how he or she intends to plead. In the case of an organization, the plea, as the whole trial appearance, may be conducted by a senior officer or representative or the retained lawyer, as is permitted by the *Criminal Code*.[65]

If the accused pleads guilty, as is done in the majority of criminal cases, he or she is admitting all the elements of the offence with which he or she is charged. Given the serious consequences arising from a criminal conviction, the accused must appreciate what he or she is pleading guilty to, with no misunderstanding. A plea of guilt must be voluntary, informed and unequivocal. If the court has some doubt whether the plea is clearly understood or unqualified, it must refuse to accept it or make inquiries to satisfy itself that the qualification or condition does not take away from the accused's intention to enter an unequivocal plea of guilty. If, on the other hand, the accused pleads not guilty, the court will proceed with a trial.[66]

At the beginning of the trial, after the jury has been selected, the trial judge will give some opening remarks to the jury informing them of the procedure that will be followed. It is then incumbent on the prosecution to give an opening statement and call witnesses to prove every fact and circumstance constituting the offence, beyond any reasonable doubt, and tender other evidence and respond to any defence that reasonably arises on the evidence. The Crown's opening statement will outline the case it

---

[64] *R. v. CIP Inc.*, [1992] I.S.C.R. 843 at 863.
[65] *Criminal Code.*, ss. 556(1) and 620.
[66] *Ibid.*, ss. 622 and 623.

intends to present, the purpose of which is to allow the trier of fact[67] to follow the evidence that will be adduced.[68] This opening address should not be used by the Crown to argue the case. The prosecution should not refer to any evidence whose admissibility must be determined.[69] Traditionally, the accused usually had no right to make an opening statement immediately after the Crown. Nevertheless, in recent years, defence counsel have been permitted to address the trier of fact in a limited fashion, if only to say a little about their case. Most defence lawyers prefer to wait until the end of the Crown's case to address the trier of fact.

After the opening address, the prosecutor begins to present evidence through the calling of witnesses. Crown counsel completes an examination-in-chief of the witness following which the defence has the right to cross-examine the witness. This defence right of cross-examination is not solely limited to issues raised during examination-in-chief, but may relate to any question which is directly in issue or which relates to the credibility of the witness. Following cross-examination, the Crown has a limited right of re-examination. It is limited because the Crown cannot raise any new areas of evidence with the witness. Rather, he or she can only ask questions relating to matters arising directly out of the cross-examination.

Once the Crown concludes its case, the accused may, if he or she desires, deliver an opening statement[70] and present evidence of its own to the trier of fact. This is done through legal counsel for the defence. Like the Crown did previously, the defence will call witnesses that will be examined-in-chief. The prosecution is then entitled to cross-examine the witness. Subject to stringent exceptions, the prosecution cannot usually cross-examine the accused on other acts of misconduct or discreditable

---

[67] This refers to the judge or the jury, whomever is responsible to make findings of fact, pursuant to the accused's elected mode of trial.

[68] Crown counsel will usually only give an opening statement during a jury trial. Nevertheless, he or she may request an opportunity to do so during a judge alone trial. Other times, the judge may request an opening summary to assist him or her in following the evidence.

[69] This will be done during a *voir dire*, or trial within a trial, whereby the Judge, in the absence of the jury, will hear arguments to determine the admissibility of evidence.

[70] *Criminal Code*, s. 651(2). The accused may also make an application for a directed verdict of acquittal on the basis that the Crown has failed to establish a sufficient case on which the trier of fact could convict him or her. The test to be applied is whether there is enough admissible evidence which could, if believed, result in a conviction. If there is, the trial judge is not justified in directing a verdict of acquittal and the defence will, if it so desires, adduce evidence of its own.

conduct for the purpose of showing that by reason of the accused's bad character, he or she ought not to be believed or that he or she is likely to have committed the offence in question.[71] Defence counsel may then re-examine the witness, subject to the same restrictions the Crown was subject to during its right of re-examination.

Following the defence's case, the prosecution may be afforded, in limited instances, the right to adduce rebuttal evidence. Generally speaking, such evidence is only admissible if new matters that could not have been anticipated were raised by the accused. More commonly, though, the judge will invite counsel to give closing statements.[72] Where an accused has elected to call evidence in defence, he or she must address the trier of fact first before the Crown. It is only if the accused elects not to call a defence that the prosecutor is required to address the trier of fact first.

After the closing statements are completed, a judge will instruct the jury, if there is one, on its role and decision. This will include instructions concerning the onus of proof, burden of proof, the presumption of innocence of the accused, the requirement that there be a unanimous verdict, the law as it relates to the specific offence or offences in question, including the elements of the crime as well as its defences. Once this charge to the jury is complete, the jury is sequestered until they reach a unanimous verdict or are unable to agree, in which case a mistrial[73] must be declared.

If the trier of fact, either a jury or a judge, finds the accused guilty, or if the accused has at any time pleaded guilty, the matter will proceed to sentencing. If the trier of fact finds the accused not guilty, the accused is acquitted and, subject to Crown appeal, the criminal prosecution is over.

SENTENCING

Sentencing occurs after an accused is found guilty and convicted of a criminal offence. Canadian sentencing jurisprudence has traditionally focused on the aims of denunciation, deterrence, separation and rehabilitation.[74] There is also a goal of restorative justice that is concerned with the restoration of the parties affected by the commission

---

[71] *R. v. McNamara (No. 1)* (1981), 56 C.C.C. (2d) 193 (Ont. C.A.).
[72] *Criminal Code*, s. 651.
[73] *Ibid.*, s. 653.
[74] Law Society of Upper Canada, "Sentencing" in Law Society of Upper Canada, Ontario Bar Admission Course Reference Materials, *Criminal Law* (Toronto: Law Society of Upper Canada, 2003) at 11-1.

of an offence. This is accomplished through rehabilitation of the offender, reparations to the victim(s) and to the community, the promotion of a sense of responsibility of the offender, and acknowledgement of the harm done. These trends have been recognized and enhanced with the Bill C-45 amendments to the sentencing provisions for organizations in the *Criminal Code*.

Offences find their associated punishment prescribed in the *Criminal Code*. An offender will thus be subject to the prescribed penalty. This includes imprisonment, fines and probation orders. Nonetheless, he or she may, if it is shown to be in his or her best interests and not contrary to the public interest, instead of being convicted, be discharged absolutely or on other conditions attached to a probation order.[75]

Organizations cannot be incarcerated. Directors and officers may not be jailed for the offence of the corporation if they have not been convicted personally. The new Bill C-45 sentencing powers for organizations, described above, are quite extensive and establish new powers for the sentencing judge. A court must take into consideration the following factors when imposing a sentence on an organization: (a) any advantage realized by the organization as a result of the offence; (b) the degree of planning involved in carrying out the offence and the duration and complexity of the offence; (c) whether the organization has attempted to conceal its assets, or convert them, in order to show that it is not able to pay a fine or make restitution; (d) the impact that the sentence would have on the economic viability of the organization and the continued employment of its employees; (e) the cost to public authorities of the investigation and prosecution of the offence; (f) any regulatory penalty that has been imposed on the organization in respect of the offence; (g) whether the organization has been convicted of a similar offence in the past; (h) any penalty that has been imposed by the organization on one of its representatives for their role in the commission of the offence; (i) any restitution that the organization is ordered to make; (j) any measures that the organization has taken to reduce the likelihood of it committing a subsequent offence.[76]

The court will be able to impose a fine on an organization,[77] make an order compelling the organization to make restitution to the victims,

---

[75] *Criminal Code*, s. 732(1)(2).
[76] Bill C-45, 2003, c. 21, s. 14.
[77] *Criminal Code*, s. 735(1). The amount of the fine is in the discretion of the court when the offence is an indictable offence. When the offence is a summary conviction offence, the imposed fine is not to exceed $100,000 (as proposed by Bill C-45).

compel the organization to establish OHS policies, standards and procedures to reduce the likelihood of the organization committing a subsequent offence and force the organization to disclose to the public information regarding the offence of which it has been convicted, the sentence it has received and any measures it is taking to prevent a recurrence of the offence.[78]

SOME FINAL THOUGHTS

The use of criminal prosecution to enforce the OHS standard is a bold new experiment in Canadian law. Putting individuals at risk for life imprisonment and organizations at risk for multi-million dollar fines underscores the importance of workplace safety in Canada. However, Bill C-45 also raises serious questions about whether criminal prosecution is the best means by which to prevent bodily harm in the workplace to workers and members of the public.

The seriousness of being charged with a criminal offence is balanced, in Canadian law, with the right given to suspects and accused under the *Charter*. The greater the risk of imprisonment and serious fines, the more rights we tend to give to suspects and accused. Therefore, individuals who may be investigated for contravention of the new offence of capital OHS criminal negligence will have their legal rights under the *Charter* recognized and upheld by the courts more than those individuals charged with OHS regulatory offences. Criminal prosecutions, rights under the *Charter,* the presumption of innocence, and jail terms will all be the new language of OHS law in Canada as a result of Bill C-45. The question will remain, will the criminalization of OHS law in Canada truly improve workplace health and safety?

Finally, it is the writer's sincere hope that the changes to the *Criminal Code* under Bill C-45 may have the effect of bringing greater attention and importance to OHS protection in every workplace across Canada. Truly, our nation's future is in the hands of our workers, especially young workers. The Bill C-45 amendments to the *Criminal Code* should give pause to all organizational decision makers, who can positively impact workplace health and safety, to double their reference to ensure that all workers, especially young workers, have a healthy and safe workplace to make a contribution to their employer's interests and to Canadian society as a whole. The development of an effective OHS

---

[78] Bill C-45, 2003, c. 21, s. 18

management system is a hallmark of the due diligence defence for OHS regulatory offences. The OHS management system is also, undoubtedly, an effective defence to a charge under the *Criminal Code* of OHS criminal negligence. More important, an effective OHS management system will prevent accidents, injury and death in the workplace.

APPENDIX A

BILL C-45

An Act to amend the *Criminal Code*
(criminal liability of organizations)
(2003, c. 21)

ASSENTED TO 7TH NOVEMBER, 2003

SUMMARY

This enactment amends the *Criminal Code* to

(a)   establish rules for attributing to organizations, including corporations, criminal liability for the acts of their representatives;

(b)   establish a legal duty for all persons directing work to take reasonable steps to ensure the safety of workers and the public;

(c)   set out factors for courts to consider when sentencing an organization; and

(d)   provide optional conditions of probation that a court may impose on an organization.

Her Majesty, by and with the advice and consent of the Senate and House of Commons of Canada, enacts as follows:

CRIMINAL CODE

1.(1) The definition "every one", "person", "owner" in section 2 of the *Criminal Code* is replaced by the following:

"every one", "person" and "owner"
"every one", "person" and "owner", and similar expressions, include Her Majesty and an organization;

(2) Section 2 of the Act is amended by adding the following in alphabetical order:

"organization"

"organization" means

(a)   a public body, body corporate, society, company, firm, partnership, trade union or municipality, or

(b)   an association of persons that

(i)   is created for a common purpose,

(ii)   has an operational structure, and

(iii)   holds itself out to the public as an association of persons:

"representative"

"representative", in respect of an organization, means a director, partner, employee, member, agent or contractor of the organization;

"senior officer"

"senior officer" means a representative who plays an important role in the establishment of an organization's policies or is responsible for managing an important aspect of the organization's activities and, in the case of a body corporate, includes a director, its chief executive officer and its chief financial officer;

2. The Act is amended by adding the following after section 22:

Offences of negligence—organizations

22.1 In respect of an offence that requires the prosecution to prove negligence, an organization is a party to the offence if

(a)   acting within the scope of their authority

(i)   one of its representatives is a party to the offence, or

(ii)   to or more of its representatives engage in conduct, whether by act or omission, such that, if it had been the conduct of only one representative, that representative would have been a party to the offence; and

(b)   the senior officer who is responsible for the aspect of the organization's activities that is relevant to the offence departs—or the senior officers, collectively, depart—markedly from the standard of care that, in the circumstances, could reasonably be expected to prevent a representative of the organization from being a party to the offence.

Other offences—organizations

22.2 In respect of an offence that requires the prosecution to prove fault—other than negligence—an organization is a party to the offence if, with the intent at least in part to benefit the organization, one of its senior officers

(a)   acting within the scope of their authority, is a party to the offence;

(b)   having the mental state required to be a party to the offence and acting within the scope of their authority, directs the work of other representatives of the organization so that they do the act or make the omission specified in the offence; or

(c)   knowing that a representative of the organization is or about to be a party to the offence, does not take all reasonable measures to stop them from being a party to the offence.

3. The Act is amended by adding the following after section 217:

Duty of persons directing work
217.1 Every one who undertakes, or has the authority, to direct how another person does work or performs a task is under a legal duty to take reasonable steps to prevent bodily harm to that person, or any other person, arising from that work or task.

4. Paragraph 328(e) of the Act is replaced by the following:

(e)   by the representatives of an organization from the organization.

5. (1) The portion of paragraph 362(1)(c) of the Act before subparagraph (i) is replaced by the following:

(c)   knowingly makes or causes to be made, directly or indirectly, a false statement in writing with intent that it should be relied on, with respect to the financial condition or means or ability to pay of himself or herself or any person or organization that he or she is interested in or that he or she acts for, for the purpose of procuring, in any form whatever, whether for his or her benefit or the benefit of that person or organization,

(2) Paragraph 362(1)(d) of the Act is replaced by the following:

(d)   knowing that a false statement in writing has been made with respect to the financial condition or means or ability to pay of himself or herself or another person or organization that he or she is interested in or that he or she acts for, procures on the faith of that statement, whether for his or her benefit or for the benefit of that person or organization, anything mentioned in subparagraphs (c)(i) to (vi).

6. Section 391 of the Act is repealed.

6.1 The portion of subsection 418(2) of the Act before paragraph (a) is replaced by the following:

Offences by representatives
    (2) Every one who, being a representative of an organization that commits, by fraud, an offence under subsection (1),

7. Paragraph 462.38(3)(b) of the Act is replaced by the following:

(b)   a warrant for the arrest of the person or a summons in respect of an organization has been issued in relation to that information, and

8. Section 538 of the Act is replaced by the following:

Organization
538. Where an accused is an organization, subsections 556(1) and (2) apply with such modifications as the circumstances require.

9. Section 556 of the Act is replaced by the following:

Organization
556. (1) An accused organization shall appear by counsel or agent.

Non-appearance
    (2) Where an accused organization does not appear pursuant to a summons and service of the summons on the organization is proved, the provincial court judge or, in Nunavut, the judge of the Nunavut Court of Justice.
(a)   may, if the charge is one over which the judge has absolute jurisdiction, proceed with the trial of the charge in the absence of the accused organization; and
(b)   shall, if the charge is not one over which the judge has absolute jurisdiction, hold a preliminary inquiry in accordance with Part XVIII in the absence of the accused organization.

Organization and electing
    (3) If an accused organization appears but does not elect when put to an election under subsection 536(2) or 536.1(2), the provincial court

judge or judge of the Nunavut Court of Justice shall hold a preliminary inquiry in accordance with Part XVIII.

10. Subsection 570(5) of the Act is replaced by the following:

Warrant of committal

(5) Where an accused other than an organization is convicted, the judge or provincial court judge, as the case may be, shall issue or cause to be issued a warrant of committal in Form 21, and section 528 applies in respect of a warrant of committal issued under this subsection.

11. The heading before section 620 and sections 620 to 623 of the Act are replaced by the following:

Organizations

Appearance by attorney
620. Every organization against which an indictment is filed shall appear and plead by counsel or agent.

Notice to organization
621. (1) The clerk of the court or the prosecutor may, where an indictment is filed against an organization, cause a notice of the indictment to be served on the organization.

Contents of notice
(2) A notice of an indictment referred to in subsection (1) shall set out the nature and purport of the indictment and advise that, unless the organization appears on the date set out in the notice or the date fixed under subsection 548(2.1), and enters a plea, a plea of not guilty will be entered for the accused by the court, and that the trial of the indictment will be proceeded with as though the organization had appeared and pleaded.

Procedure on default of appearance
622. Where an organization does not appear in accordance with the notice referred to in section 621, the presiding judge may, on proof of service of the notice, order the clerk of the court to enter a plea of not guilty on behalf of the organization, and the plea has the same force and effect as if the organization had appeared by its counsel or agent and pleaded that plea.

Trial of organization

623. Where an organization appears and pleads to an indictment or a plea of not guilty is entered by order of the court under section 622, the court shall proceed with the trial of the indictment and, where the organization is convicted, section 735 applies.

12. Subsection 650(1) of the Act is replaced by the following:

Accused to be present

650. (1) Subject to subsections (1.1) to (2) and section 650.01, an accused, other than an organization, shall be present in court during the whole of his or her trial.

13. Section 703.2 of the Act is replaced by the following:

Service of process on an organization

703.2 Where any summons, notice or other process is required to be or may be served on an organization, and no other method of service is provided, service may be effected by delivery.

(a)   in the case of a municipality, to the mayor, warden, reeve or other chief officer of the municipality, or to the secretary, treasurer or clerk of the municipality; and

(b)   in the case of any other organization, to the manager, secretary or other senior officer of the organization or one of its branches.

14. The Act is amended by adding the following after section 718.2:

Organizations

Additional factors

718.21 A court that imposes a sentence on an organization shall also take into consideration the following factors:

(a)   any advantage realized by the organization as a result of the offence;

(b)   the degree of planning involved in carrying out the offence and the duration and complexity of the offence;

(c)   whether the organization has attempted to conceal its assets, or convert them, in order to show that it is not able to pay a fine or make restitution;

(d)   the impact that the sentence would have on the economic viability of the organization and the continued employment of its employees;

(e)   the cost to public authorities of the investigation and prosecution of the offence;

(f)   any regulatory penalty imposed on the organization or one of its representatives in respect of the conduct that formed the basis of the offence;

(g)   whether the organization was—or any of its representatives who were involved in the commission of the offence were—convicted of a similar offence or sanctioned by a regulatory body for similar conduct;

(h)   any penalty imposed by the organization on a representative for their role in the commission of the offence;

(i)   any restitution that the organization is ordered to make or any amount that the organization has paid to a victim of the offence; and

(j)   any measures that the organization has taken to reduce the likelihood of it committing a subsequent offence.

15. Subsection 721(1) of the Act is replaced by the following:

Report by probation officer
721. (1) Subject to regulations made under subsection (2), where an accused, other than an organization, pleads guilty to or is found guilty of an offence, a probation officer shall, if required to do so by a court, prepare and file with the court a report in writing relating to the accused for the purpose of assisting the court in imposing a sentence or in determining whether the accused should be discharged under section 730.

16. Subsection 727(4) of the Act is replaced by the following:

Organizations
(4) If, under section 623, the court proceeds with the trial of an organization that has not appeared and pleaded and convicts the organization, the court may, whether or not the organization was notified that a greater punishment would be sought by reason of a previous conviction, make inquiries and hear evidence with respect to previous convictions of the organization and, if any such conviction is proved, may impose a greater punishment by reason of that conviction.

17. Subsection 730(1) of the Act is replaced by the following:

Conditional and absolute discharge
730. (1) Where an accused, other than an organization, pleads guilty to or is found guilty of an offence, other than an offence for which a minimum

punishment is prescribed by law or an offence punishable by imprisonment for fourteen years or for life, the court before which the accused
appears may, if it considers it to be in the best interests of the accused
and not contrary to the public interest, instead of convicting the accused,
by order direct that the accused be discharged absolutely or on the conditions prescribed in a probation order made under subsection 731(2).

**18. (1) The definition "optional conditions" in subsection 732.1(1) of
the Act is replaced by the following:**

"optional conditions"
"optional conditions" means the conditions referred to in subsection (3)
or (3.1).

(2) Section 732.1 of the Act is amended by adding the following after
subsection (3):

Optional conditions—organization
(3.1) The court may prescribe, as additional conditions of a probation
order made in respect of an organization, that the offender do one or
more of the following
(a)   make restitution to a person for any loss or damage that they
      suffered as a result of the offence;
(b)   establish policies, standards and procedures to reduce the
      likelihood of the organization committing a subsequent offence;
(c)   communicate those policies, standards and procedures to its
      representatives;
(d)   report to the court on the implementation of those policies,
      standards and procedures;
(e)   identify the senior officer who is responsible for compliance with
      those policies, standards and procedures;
(f)   provide, in the manner specified by the court, the following
      information to the public, namely,
      (i)   the offence of which the organization was convicted,
      (ii)   the sentence imposed by the court, and
      (iii) any measures that the organization is taking—including
             any policies, standards and procedures established under
             paragraph (b)—to reduce the likelihood of it committing a
             subsequent offence; and
(g)   comply with any other reasonable conditions that the court
      considers desirable to prevent the organization from committing
      subsequent offences or to remedy the harm caused by the offence.

Consideration—organizations

(3.2) Before making an order under paragraph (3.1)(b), a court shall consider whether it would be more appropriate for another regulatory body to supervise the development or implementation of the policies, standards and procedures referred to in that paragraph.

19. The portion of subsection 734(1) of the Act before paragraph (a) is replaced by the following:

Power of court to impose fine

734. (1) Subject to subsection (2), a court that convicts a person, other than an organization, of an offence may fine the offender by making an order under section 734.1.

20. (1) The portion of subsection 735(1) of the Act before paragraph (a) is replaced by the following:

Fines on organizations

735. (1) An organization that is convicted of an offence is liable, in lieu of any imprisonment that is prescribed as punishment for that offence, to be fined in an amount, except where otherwise provided by law,

(2) Paragraph 735(1)(b) of the Act is replaced by the following:

(b) not exceeding one hundred thousand dollars, where the offence is a summary conviction offence.

(3) Subsection 735(2) of the Act is replaced by the following:

Effect of filing order

(2) Section 734.6 applies, with any modifications that are required, when an organization fails to pay the fine in accordance with the terms of the order.

21. Subsection 800(3) of the Act is replaced by the following:

Appearance by organization

(3) Where the defendant is an organization, it shall appear by counsel or agent and, if it does not appear, the summary conviction court may, on proof of service of the summons, proceed *ex parte* to hold the trial.

COORDINATING AMENDMENT

22. On the later of the coming into force of section 9 of this Act and section 34 of the *Criminal Law Amendment Act, 2001,* section 556 of the *Criminal Code* is replaced by the following:

Organization

556. (1) An accused organization shall appear by counsel or agent.

Non-appearance

(2)   Where an accused organization does not appear pursuant to a summons and service of the summons on the organization is proved, the provincial court judge or, in Nunavut, the judge of the Nunavut Court of Justice.

(a)   may, if the charge is one over which the judge has absolute juris-diction, proceed with the trial of the charge in the absence of the accused organization; and

(b)   shall, if the charge is not one over which the judge has absolute jurisdiction, fix the date for the trial or the date on which the accused organization must appear in the trial court to have that date fixed.

Preliminary Inquiry not requested

(3) If an accused organization appears a preliminary inquiry is not re-quested under subsection 536(4), the provincial court judge shall fix the date for the trial or the date on which the organization must appear in the trial court to have that date fixed.

Preliminary Inquiry not requested—Nunavut

(4)   If an accused organization appears and a preliminary inquiry is not requested under subsection 536.1(3), the justice of the peace or the judge of the Nunavut Court of Justice shall fix the date for the trial or the date on which the organization must appear in the trial court to have that date fixed.

Order

23. The provisions of this Act, other than section 22, come into force on a day or days to be fixed by order of the Governor in Council.

APPENDIX B

WESTRAY MINE DISASTER FACTUAL AND LEGAL CHRONOLOGY

Date	Event or Legal Developments
May 9, 1992	Westray mine in Plymouth, Pictou County, Nova Scotia explodes. Twenty-six miners died in the mine. Only fifteen bodies were eventually recovered.
May 10, 1992	A Public Inquiry is promised by the province of Nova Scotia.
May 15, 1992	Premier Donald Cameron appoints Mr. Justice Peter Richard of the Nova Scotia Supreme Court to carry out the Westray Public Inquiry.
May 21, 1992	The RCMP launch criminal investigation into the Westray disaster, in part, due to allegations that documents were being shredded at the mine.
Sept. 30, 1992	*Phillips v. Nova Scotia (Commission of Inquiry into the Westray Mine Tragedy):*[1] An application for an interim injunction of the Public Inquiry was brought by management employees at Westray mine. Chief Justice Constance Glube of the Nova Scotia Supreme Court issues a temporary injunction halting the Public Inquiry to be held by Justice Peter Richard, in his capacity as Commissioner. They sought to have the Public Inquiry suspended until the determination of the constitutional validity of the Public Inquiry

---

[1] (1992), 116 N.S.R. (2d) 30 (N.S.S.C.).

Date	*Event or Legal Developments*
	and in light of potential criminal charges being laid against the same individuals. The interim injunction was granted. Applying the test for issuing injunctions, Chief Justice Glube held that (a) there was a serious issue to be tried; (b) the applicants would suffer irreparable harm that would be difficult to compensate in damages; and (c) despite the delay, the public and private interests were best served by granting an interim injunction, since the Public Inquiry, if it proceeds, will be pronounced constitutionally valid.
October 5, 1992	Nova Scotia Department of Labour lays 52 charges under the *Occupational Heath and Safety Act*, R.S.N.S. 1989, c. 320 ("OHSA") and the *Coal Mines Regulation Act*, R.S.N.S. 1989, c. 73.
November 1992	Judicial hearings held before Chief Justice Glube. Public Inquiry representatives express an intention to avoid naming names in their inquiry. The government of Nova Scotia lawyers supports a ban on all hearings until the prosecutions are complete.
November 13, 1992	*Phillips v. Nova Scotia (Commission of Inquiry into the Westray Mine Tragedy)*:[2] Individual employees and management personnel of Westray mine brought an application against Justice Peter Richard to have the Public Inquiry, which was created by an Order in Council, prohibited from carrying out any of its objectives. Chief Justice Glube strikes down the Inquiry as being *ultra vires*, or beyond the jurisdiction of the province. Specifically, the applicants sought to have the Public Inquiry declared *ultra vires* the province on the bases that (a) the scope of powers granted encroach on criminal law powers that fall within federal

---

[2]   (1993), 116 N.S.R. (2d) 34 (N.S.S.C.).

Date	*Event or Legal Developments*
	jurisdiction; and (b) that the powers conferred are those rightfully exercised by section 96 courts, and thus derogate from the power of the Governor General to appoint section 96 judges. Further, the applicants sought a declaration that the Order in Council establishing the Public Inquiry violated rights of the applicants guaranteed by the *Charter*. Chief Justice Glube declared that the Public Inquiry was *ultra vires* the province since the terms of reference of the Westray Public Inquiry were effectively a criminal investigation and the dominant purpose of the Public Inquiry was to assign probable criminal, quasi-criminal and civil responsibility. The applications relating to section 96 of the *Constitution Act* and the *Charter* were dismissed.
December 1992	The 34 Department of Labour OHSA charges were voluntarily stayed, or dropped, including serious allegations such as tampering with methane detectors and failing to prevent build-up of coal dust.
January 19, 1993	*Phillips v. Nova Scotia (Westray Mine Public Inquiry):*[3] The Attorney General and others appealed the decision of Glube C.J. declaring the Order in Council constituting the Public Inquiry, and related statutory provisions *ultra vires* the province of Nova Scotia. The appeal is allowed. The Nova Scotia Court of Appeal unanimously reinstates the Public Inquiry, finding it within the province's lawful powers. However, a stay is issued on the public hearings until both the criminal charges, and the charges under the *Occupational Health and Safety Act*, are disposed of or stayed. The court found the Order in Council creating the Public Inquiry as well as the impugned provisions of

---

[3] (1993), 117 N.S.R. (2d) 218 (N.S.C.A.).

Date	*Event or Legal Developments*
	the *Coal Mines Regulation Act* to be *intra vires* or within the jurisdiction of the province. However, Hallett J.A. opined that both the right to silence, embodied in section 7, and the right to a fair trial under section 11(d), of the *Charter*, would be infringed if the Commissioner compelled the respondents charged with quasi-criminal offences to testify before him. The court held that this would prevail as long as the charges against them under the *Occupational Health and Safety Act* as were alive and the criminal investigation was ongoing. Therefore, the court ordered the Public Inquiry to postpone its public hearings until the charges against the four respondents under the *Criminal Code* or the *Occupational Health and Safety Act* are disposed of or stayed.
February 22, 1993	Both of the provincial prosecutors assigned to the Westray criminal investigations, at the new Office of Public Prosecution, resign for reasons given of insufficient funding and administrative support to continue their duties.
March 1993	The outstanding Department of Labour charges under the OHSA were stayed to ensure that the RCMP criminal investigation continue and to prevent the possibility of a court rejecting the criminal charges on the basis of double jeopardy or other *Charter* rights.
March 26, 1993	Lawyers for Curragh ask the court to order the return of various documents and evidence that were in the possession of the RCMP without any charges being laid. The judge gives the RCMP one month to either lay criminal charges or return the evidence to Curragh.
April 20,1993	Curragh, Gerald Phillips, mine manager, and Roger Parry, the Westray underground manager, are formally charged with man-slaughter and criminal negligence causing

Date	*Event or Legal Developments*
	death under the *Criminal Code* in relation to the Westray mine disaster.
July 20, 1993	*R. v. Curragh Inc.*:[4] The accused brought a motion to quash the Information laying the criminal charges on the grounds of both legal and factual deficiencies. Provincial Court Judge Patrick Curran quashes the criminal charges for vagueness. Judge Curran held that the information on both counts to be fundamentally deficient in that the charge gives no indication whatever of the ways the coal mine was allegedly operated so as to constitute the criminal offence[s]. The alleged manslaughter and criminal negligence were based on a broad and unspecified range of possible acts or omissions that were completely devoid of particulars.
October 26, 1993	*R. v. Curragh Inc. (No. 2)*:[5] After new charges are laid against Curragh, Phillips, and Parry, counsel for the accused raised in eight separate arguments in support of quashing the second Information. This motion to argue that the second set of criminal charges were defective is rejected in the Provincial Court. Judge Curran allowed the Information to stand and dismissed the motion to quash finding that both counts were valid as laid.
November 17, 1993	*R. v. Curragh Inc.*:[6] The Crown appealed the court order to return items seized in a search to Curragh. Appeal from trial judge's decision to order the return of items seized under search warrants is dismissed. The Crown had requested continued detention of items such as coal dust samples, photographs and mining equipment.

---

[4] (1995), 146 N.S.R. (2d) 163 (N.S.S.C.), rev'd (1995), 146 N.S.R. (2d) 161, aff'd (1997), 113 C.C.C. (3d) 481 (S.C.C.).
[5] (1993), 125 N.S.R. (2d) 185 (N.S. Prov. Ct.).
[6] (1993), 126 N.S.R. (2d) 159 (N.S.S.C.).

Date	*Event or Legal Developments*
	The request was denied in Provincial Court. The trial judge's finding that the time for applying for the detention order had expired was correct. Justice MacLellan found that it was not appropriate for the court to consider the fact that criminal charges were laid after the order made for return of items, as at the time this order was made there was a finding that no criminal charges had been instituted and therefore, the items were properly ordered returned.
February 1995	Criminal trial begins in Pictou, Nova Scotia, lasting 44 days. A total of 23 witnesses are called to testify. The proceedings focus primarily on issues relating to Crown disclosure rather than the evidence respecting adequacy of health and safety measures and causes of the disaster at Westray.
March 2, 1995	Justice Robert Anderson makes a telephone call to Crown Attorney's office demanding that the lead prosecutor be taken off the case. Crown prosecutors learn of this telephone call by the trial Judge and demand a mistrial. Anderson refuses to grant a mistrial.
April 5, 1995	*R. v. Curragh Inc.*:[7] In response to Justice Anderson's telephone call, the Crown had brought a motion for a mistrial, which was dismissed by Judge Anderson. At an emergency hearing held at the Supreme Court of Canada to seek the court's intervention relating to the conduct of the trial judge, Chief Justice Lamer refuses to intercede in the proceedings. In a brief decision, Chief Justice Lamer states that the Supreme Court of Canada had no jurisdiction to hear an appeal at this stage of the proceedings and therefore, refrained from expressing any views as regards the merits of the matter.

---

[7] [1995] 1 S.C.R. 900.

Date	Event or Legal Developments

Date

Event or Legal Developments

May 4, 1995

Phillips v. Nova Scotia (Commission of Public Inquiry into the Westray Mine Tragedy):[8] The union representing the miners of Westray appealed the decision of the Court of Appeal staying the public hearings pending disposition of all other charges at trial. Supreme Court of Canada grants the appeal by the United Steelworkers of America, Local 9332 and lifts the ban on the Public Inquiry's hearings. In three separate judgments, the court allowed the appeal and set aside the stay on the public hearings. Five Justices held that the Charter issue was no longer alive, since the accused had subsequently changed its election and elected to proceed without a jury trial. The same Justices found the issue regarding compellability of the witnesses at the Public Inquiry to be premature and refused to pronounce on either issue for future guidance. The other four Justices believed the witnesses were compellable to testify at the Inquiry, and that their right to a fair trial had not been infringed by holding simultaneous proceedings. The court considered a stay of proceedings to be an excessive remedy in the circumstances.

June 9, 1995

On a defence motion, trial judge Justice Robert Anderson rules that the rights of the accused under section 7 of the *Charter* was violated as a result of non-disclosure and late disclosure. The evidence in question was held to be essential to a fair trial as well as the ability of the accused to make full answer and defence. Justice Anderson held that the accused were not treated fairly and that the fundamental principles of justice were not observed. An award of costs was also granted against the Crown.

[8] (1995), 124 D.L.R. (4th) 129 (S.C.C.).

Date	*Event or Legal Developments*
December 1, 1995	*R. v. Curragh Inc.*:[9] Crown appealed Judge Anderson's decision to stay the proceedings on the basis of non-disclosure or late disclosure of evidence. The appeal is granted. A new trial is ordered in the Court of Appeal by Chief Justice Hallett.
	The learned trial judge was held to have done the following: (a) demonstrated an appearance of bias that affects his decision on the stay (b) he failed to make an inquiry with respect to whether evidence that had not been disclosed or was disclosed late was material and prejudicial to the respondents' ability to make full answer and defence; and (c) he failed to exercise his power to grant a stay in a judicial manner.
March 15, 1996	*Stellarton (Town) v. Nova Scotia (Commission of Public Inquiry into the Westray Mine Tragedy)*:[10] The Town of Stellarton brought this application after it was denied funding required to participate in the Public Inquiry. Town seeks Writ of Certiorari directing the Commissioner to review his decision declining funding to participate in the Public Inquiry. The Nova Scotia Supreme Court dismissed the application. Counsel alleged bias or reasonable apprehension of bias on the part of the Commissioner. The application was dismissed. Justice Goodfellow held that the Commissioner's exercise of discretion was appropriate and there was no evidence or conduct establishing bias or a reasonable apprehension of bias.

---

[9] (1995), 146 N.S.R. (2d) 163 (N.S.S.C.), rev'd (1995), 146 N.S.R. (2d) 161, aff'd (1997), 113 C.C.C. (3d) 481 (S.C.C.).
[10] (1996), 150 N.S.R. (2d) 11 (N.S.S.C.).

Date	Event or Legal Developments
November 19, 1996	*Frame v. Nova Scotia (Commission of Public Inquiry into the Westray Mine Tragedy):*[11] Appellants applied to a chambers judge for an order requiring (i) the production of documents; and (ii) the attendance of a witness for examination for discovery. This was an appeal from the dismissal of an application to require the production of documents and attendance of a witness at an examination for discovery. This was an appeal from that decision. The appeal was dismissed. The chambers judge orally gave an unofficial decision, which is not mentioned in the Rules of Civil Procedure and…is not included in th[e] definition of 'judgment'. Thus, the court held, no appeal lies to the Court of Appeal from the document.
December 13, 1996	*Frame v. Nova Scotia (Commission of Public Inquiry into the Westray Mine Tragedy):*[12] Appeal from a dismissal of a chambers judge of applications for production of documents and examination for discovery of a witness. Appeal from a dismissal of applications for the production of documents and for the discovery examination of a witness. The appeal was dismissed. As the decision of the chambers judge did not result in a foreclosure of the issues between the parties, there was no legal error requiring intervention from the Court of Appeal. Similarly, there was no evidence of the application of wrong principles of law or of patent injustice by the chambers judge.
March 20, 1997	*Phillips and Parry v. The Queen*[13] (also indexed as: *R. v. Curragh Inc.*): After the Court of Appeal set aside the stay of

[11] (1996), 155 N.S.R. (2d) 238 (N.S.C.A.).
[12] (1996), 157 N.S.R. (2d) 77 (N.S.C.A.).
[13] (1995), 146 N.S.R. (2d) 163 (N.S.S.C.), rev'd (1995), 146 N.S.R. (2d) 161, aff'd (1997), 113 C.C.C. (3d) 481 (S.C.C.).

Date	*Event or Legal Developments*
	proceedings and ordered a new trial in the prosecution of Curragh, Phillips and Parry, the accused brought this appeal to uphold the trial judge's stay to the Supreme Court of Canada. In a split decision (7:2), the Supreme Court of Canada agrees with the Nova Scotia Court of Appeal that there should be a new trial. Two justices, McLachlin and Major JJ., strongly dissent. The Court of Appeal was found to have correctly held that the trial was unfair as a result of the evidence of a reasonable apprehension of bias on the part of the trial judge. Therefore, the order of the trial judge staying the charges for failure to provide full, complete and timely disclosure was void. The Court issued a somewhat unusual order for costs against the Crown, demonstrating their disapproval of the egregious conduct relating to the disclosure issue, as well in recognition of the trial judge's refusal to recuse himself from the trial.
July 17, 1997	*Nova Scotia (Commission of Public Inquiry into the Westray Mine Disaster) v. Frame*:[14] The Commissioner, Justice Peter Richard, brought this application to have the Ontario Court receive and adopt subpoenas for witnesses at the public hearings of the Public Inquiry in Nova Scotia or, in the alternative, to have the witnesses appear before a Commissioner in Ontario to be examined under oath on matters relating to the Public Inquiry. The two witnesses, in question, Clifford Frame (Chairman and CEO of Curragh Inc.) and Marvin Pelley (President of Curragh Inc.), were residents of Ontario and refused to give evidence voluntarily at the Public Inquiry's hearings. The application was

---

[14] [1997] O.J. No. 5425 (Ont. Gen. Div.).

Date	*Event or Legal Developments*
	granted. The Commissioner can decide whether to proceed by subpoena in Nova Scotia or by Commission in Ontario. Justice Sheard held that (i) the certificate issued in Nova Scotia meets the requirements under the Ontario Interprovincial Summonses Act; (ii) that the Public Inquiry constitutes a court in the context of this application; and (iii) that a broad interpretation should be applied to the provision that evidence must relate to a civil, commercial or criminal matter.
September 26, 1997	*Canada (Attorney General) v. Canada (Commission of Inquiry on the Blood System):*[15] Certain individuals and organizations under investigation by the Krever Commission Inquiry on the Blood System brought this appeal regarding the jurisdiction of the Commissioner to reach conclusions of evidence that might amount to findings of misconduct. Supreme Court of Canada releases a judgment clarifying the scope of powers held by public inquiries, and permitting the assignment of individual blame in such an Inquiry. Justice Richard takes some judicial direction from this decision of the Supreme Court of Canada and expresses direct criticism of Frame and Pelley in his final report. The Supreme Court of Canada held that the Commissioner did not exceed his jurisdiction: "The Public Inquiry's mandate was broad and included investigation resulting in potential findings of misconduct."
June 30, 1998	Public announcement is made in Nova Scotia that a second trial will not proceed and that criminal charges against Phillips and Parry are stayed.

[15] [1997] 3 S.C.R. 440.

Appendix C

Consolidated Recommendations of the Report of the Westray Mine Public Inquiry, Justice K.P. Richard, Commissioner

1.   No provisional mining certificate should be issued at any circumstance. The process of granting certification based on status in other jurisdictions must be refined to ensure that qualifications are consistent with provincial requirements. The burden should be on the applicant to establish that his or her qualifications are sufficient to support the requirements for the certification sought. Any person granted certification based on status in another jurisdiction should be required to be examined in Nova Scotia for such certification at the earliest reasonable time.

2.   Every position in a mine should have a written job description setting out the duties and responsibilities of that position, with particular reference to safety. Each employee should be provided with a copy of his or her job description. A copy of all job descriptions should be prominently displayed in an area frequented by employees.

3.   One regulatory organization (such as the Department of Labour or a board of examiners) should be responsible for certifying workers in underground coal mines in Nova Scotia.

4.   Before approving the start-up of any underground coal mine, the regulator should review and amend the standards of certification to ensure the following:

    (a)   Standards of certification fit the mining methods and technology of the proposed mine.

(b)   All positions in the mining operation are filled by people with the qualifications and experience necessary to do their jobs safely.

(c)   The system of certification applies to every person required to work underground. Categories of certification should include (at a minimum) coal miner, electrical tradesperson, mechanical tradesperson, surveyor, engineer, mine rescue person, and the various levels of supervisors and managers.

(d)   Trainers have the necessary qualifications and experience.

5.   The regulator should establish a model curriculum consistent with established standards and practices in the coal mining industry.

6.   The mine operator should be required to have in place a training program, approved by the regulator, for every position in the workplace. The mine operator's training proposal must:

(a)   conform to or be more rigorous than the model curriculum;

(b)   show when, how, and what training will be done;

(c)   incorporate annual refresher training and safety education;

(d)   provide for adequate orientation to the mine for all new employees, including those with experience in coal mines; and

(e)   include complete and sufficient training for operators of individual pieces of mining equipment prior to their being assigned operating positions.

7.   The mine operator should be required to keep training and work history records for applicants for certification. The regulator should:

(a)   check applicants' records, making sure that training is taking place; and

(b)   test applicants for certification in a manner that establishes whether underground workers are trained sufficiently to work safely.

8.   The mine's joint occupational health and safety committee should periodically review training standards, policies, and programs to make sure that they adequately reflect changing technology and mining conditions and practice within the mine.

9.   Incentive bonuses based solely on productivity have no place in a hazardous working environment such as an underground coal mine. Such schemes should be replaced, where practical, by safety incentives that include three principles:

(a)   The incentive plan should be developed cooperatively with the employees to whom it will be addressed.

(b)   Both group safety performance and individual safety performance should be rewarded.

(c)   All employees, whether underground or on surface — workers, supervisors, and middle managers — should be included.

If properly instituted, such a safety incentive plan may well have its own productivity rewards.

10.   The overriding principle in mine ventilation must be that the mine is properly ventilated at all working times. It is the primary duty of the mine manager to ensure this proper ventilation.

(a)   All active working places should be ventilated by a current of fresh air containing not less than 19.5 per cent by volume of oxygen and not more than 0.5 per cent by volume of carbon dioxide.

(b)   Each working face should receive fresh air of sufficient volume and velocity to dilute and render harmless all noxious or flammable gases and maintain all working and travelling areas in a safe and fit condition.

11.   No mine should start up without a comprehensive ventilation plan approved by the regulator. The ventilation plan should be subject to at least an annual update, and any changes in the interim should be subject to approval by the regulator.

12.   The ventilation plan should contain details of the system proposed, or of amendments to the existing approved system, and should indicate:

(a)   the limits of the mine property and any adjacent workings, as well as any abnormal conditions;

(b)   the location and detailed specifications of all surface fans and all surface openings;

(c)   the direction, velocity, and volume of air at each mine opening;

(d)   all underground workings, including location of all stoppings, overcasts, undercasts, regulators, doors, and seals;

(e)   the method of sealing worked-out areas, provisions for air sampling behind any such seals, and the manner in which such sealed areas will be vented into return air passages (ensuring that no intake air is or could be passing any sealed-off area);

(f)   the location of all splits and the volume of fresh air enter-
        ing each split and of return air at each cross-cut in a room-
        and-pillar mine and at each working face; and

(g)   the locations for the measurement of air in the mine to en-
        sure the proper ventilation at all times.

13.   The mine operator should employ or retain the services of a
        qualified ventilation engineer to assist in the preparation of all
        ventilation plans or amendments to such plans. The ventilation
        engineer should sign any ventilation plans or amendments before
        they are submitted to the regulator.

14.   The regulator may submit plans or amendments to a qualified
        mine ventilation engineer for review, and any fee for such review
        should be the responsibility of the mine operator. The regulator
        may require modifications to the plan in the interests of safety.

15.   The regulator, in consultation with a qualified ventilation
        engineer, should draft regulations dealing with main fans and
        auxiliary fans. These regulations should include:

(a)   details of the design, installation, operation, maintenance,
        and inspections of such fans; and

(b)   requirements for instrumentation, the recording of data
        from such instrumentation, and the filing of this data with
        the regulator.

16.   No booster fan should be installed underground without the
        approval of the regulator.

17.   Every main ventilating fan should be mounted above ground in a
        fireproof fan house located at a safe distance from any mine
        opening and offset from any such openings or connections. The
        fan house should be equipped with a weak wall or explosion door
        located in a direct line with any possible explosion forces. Every
        main fan should be equipped with an audible alarm that sounds
        automatically if the fan stops or slows down.

18.   Where any fan used in ventilating a mine stops for any reason, the
        area affected should be immediately evacuated. No auxiliary fan
        should be restarted until a qualified person has inspected the area
        and found it to be safe and free of gas. The area should not be re-
        entered until the ventilation has been restored to the required level
        and the area has been found to be safe and free of gas by a
        qualified person. If any fan remains stopped for more than 30
        minutes, the mine operator should report the relevant
        circumstances to the regulator.

19.   The regulator, in consultation with a qualified ventilation engineer, should draft regulations dealing with requirements for ducting, brattice, stoppings, locations of measuring devices, and sealing of abandoned sections of the mine. All brattice cloth, ducting, and materials used for constructing stoppings should be of fire-resistant material.

20.   Equipment used to ventilate an underground coal mine should be of a type approved by the regulator and should be installed in an approved manner. Equipment, materials, or procedures not previously approved may be approved if the regulator is satisfied that the same measure of protection is provided to the underground worker.

21.   Unless specifically approved in writing by the regulator, no more than one mechanized coal mining unit should operate in each ventilation split. Each split should be provided with a separate supply of fresh air.

22.   Ventilating air should not be recirculated without the written consent of the regulator.

23.   The mine operator should employ a qualified mine ventilation technician to be responsible for the operation and maintenance of the ventilation system. The ventilation technician should measure the airflow and sample the air quality in the mine at approved intervals of at least once a month for the whole mine and weekly for working areas. The results of ventilation and air quality tests should be recorded and a copy of such record should be filed with the regulator.

24.   Workers should be removed from any area in a mine where the concentration of dust or noxious gases in the air exceeds the standards set out by the American Conference of Governmental Industrial Hygienists (ACGIH).

25.   Devices used for testing air quality, velocity, and volume should be of a type certified and approved for such use by the Canada Centre for Mineral and Energy Technology (CANMET), the Approval and Certification Center of the Mine Safety and Health Administration (MSHA), the Canadian Standards Association (CSA), or other such equivalent testing body.

26.   The level of methane in an air intake to the working face of the mine should not exceed 0.5 per cent by volume.

    (a)   If the methane level exceeds 0.5 per cent by volume, the ventilation technician or other qualified person must take steps to adjust the ventilation system to dilute the methane to acceptable levels.

(b) If the methane level in any part of a mine reaches or exceeds 2 per cent by volume, all workers must be evacuated from the affected area.

(c) The airflow throughout the mine, including the mine face, should be such that methane will be diluted to a level below 0.5 per cent by volume, as measured at least 30 cm from the roof or ribs.

(d) The velocity of air throughout the mine should be sufficient to prevent the formation of methane layers.

27. Each crew at the working face of a mine should include a person trained in the use of a methanometer. This person should carry, while in the mine, an approved device or devices capable of testing for both methane and oxygen, and capable of testing at the roof and in roof cavities for layering.

28. The mine operator should provide suitable testing and calibrating facilities on the mine surface. Methanometers should be tested for accuracy before each shift and calibrated as required.

29. If the locked flame safety lamp is used at all, it should be handled only by persons who have received adequate training in its assembly and operation. No lamp should be reignited underground unless the methane content in the ambient air is 0 per cent, as determined by a methanometer.

30. If the methane level in the area reaches or exceeds 1 per cent by volume, any electrically operated equipment in use should be shut down, and any shotfiring being carried out should be discontinued.

(a) In addition to other safety devices, any electrical equipment operating at the mine face or in reasonable proximity, as established by the regulator, should be equipped with a methane-monitoring device capable of continually monitoring the methane content of the air.

(b) If the methane content exceeds 1 per cent by volume, the methane monitoring device should automatically shut down the electrical equipment.

(c) The electrical equipment should not be re-energized until a qualified person certifies that the methane content in the air has been diluted to a safe level. (30 CFR sets out this requirement as it applies to mines under the jurisdiction of the U.S. Mine Safety and Health Administration.)

(d)   The methane monitors installed on electrical equipment should be kept operative at all times and tested weekly for accuracy. Sensors should be affixed to the equipment as close to the working face as practicable.

31.   The operation of mobile diesel-powered equipment underground should be regulated to ensure that the health and safety of the workforce is not endangered or impaired by such operation.

32.   The regulator may require, as part of the mine development plan, a plan for the installation of a remote system for monitoring the mine atmosphere, with appropriate audible alarms and recording devices. Such a monitoring plan should include the provision that a qualified person must be at the remote monitoring station at all times that the mine is operating.

33.   As a prerequisite to the resumption of underground coal mining at Westray or elsewhere in the Pictou coal basin, the province should require the completion of a study into the safety and economic factors involved in drainage of the coalbed methane in the mining area concerned.

34.   Every mine development plan should include complete details of any program or process designed to drain methane from the coal seam before, during, and after mining. The regulator could waive this requirement if satisfied that the program or process would be impractical and that general mine safety would not be compromised.

35.   Every coal mine operator should prepare a program for the regular cleanup and removal of coal dust and other combustibles from the floor, roof, and ribs of roadways and work areas in the mine. A copy of the program should be filed with the regulator, who may require changes in the cleanup program if it does not comply with accepted industry standards.

36.   Sufficient water should be provided in the mine to ensure that an adequate supply is available to wet the coal being mined and transported within the mine.

(a)   All coal-cutting picks should be equipped with water-spray jets of sufficient number and size to ensure that the areas of the coal face being worked are maintained in a damp condition so as to render any coal dust incombustible.

(b)   All transfer points where coal is moved from one mode of transport to another should be equipped with water-spray devices sufficient to render any coal dust incombustible.

37.   The Department of Labour and the Department of Natural Resources should consider active research in the development and use of passive and triggered stonedust and water barriers for the drives and entries of underground coal mines. This research should be aimed at the development of such techniques for use in room-and-pillar mining operations. If the development of barrier technology indicates that substantial safety benefits may accrue, the regulator could order a mine operator to install water or stonedust barriers in the mine.

38.   All underground areas of a coal mine should be stonedusted to within 12 m of the working face and all cross-cuts less than 12 m distant from the face should be stonedusted. This would not apply to those areas within the mine containing sufficient moisture to render the coal dust incombustible or for which the regulator, after examination, has granted exemption.

39.   A mine operator should file with the regulator a copy of the stonedusting program for the mine, including the method and frequency of testing; the type of testing equipment used; the type and number of dust-spreading machines used; the frequency of dusting; and the location and quantity of stonedust stored in the mine for emergencies (as opposed to normal usage).

40.   The material used for stonedusting should be of a type approved by the regulator for that purpose and should meet accepted industry standards as to size, composition, and incombustibility.

41.   Dust samples should be taken at least once a week using a method approved by the regulator for that purpose. Samples should be taken according to a regularly updated and approved plan. The regulator may require additional testing and may grant exemptions, providing that the overall safety of underground workers is not compromised.

42.   Consultants, when required, should be selected carefully to ensure that their background and expertise are consistent with the specific requirements of the problem to be analysed. Any conflicts in the advice from these consultants ought be resolved through discussion and, if necessary, through further advice. Conflicts in technical advice must be resolved, not ignored.

43.   A legislative regime should be put in place to ensure regulatory involvement in all areas of ground control in which safety is a consideration. The regime should encompass planning approval, materials and equipment certification, and any other aspect of ground control having safety implications.

44.   The regulations should specify the following at a minimum:

(a)   Ground control plans and any revisions to those plans should be prepared by the mine operator and submitted to the regulator for approval prior to the implementation of any such plans.

(b)   The ground control plan should show the existing geological conditions and the mining system to be used. The plan should also indicate any unusual hazards and outline the manner in which these will be handled.

(c)   Approved plans should be available to miners and other underground workers and should be posted in the mine at the area affected by the plan.

(d)   What the plan is required to specify should be set forth by the regulator from time to time, and should include:

(i)   a columnar section of mine strata;

(ii)   planned width of openings and size of pillar (if required);

(iii)   thickness of seam;

(iv)   method of support to be used;

(v)   type, sequence, and spacing of support materials;

(vi)   requirements for temporary roof support systems; and

(vii)   type and thickness of strata in the roof and in the floor for a depth of 3 m below the coal bed.

(e)   The regulator may require further and better information on the plan and may require that the plan be reviewed by a qualified specialist in rock mechanics.

(f)   The regulator may require revisions to the plan at any time if satisfied that conditions or accident experience indicate that such revisions are necessary or conducive to safety.

(g)   The ground control plan should be reviewed at least once every six months by the regulator.

(h)   The mine operator should record on the plan and report to the regulator any unplanned fall of roof or rib or any significant rock burst (more than 0.3 m in thickness) that occurs above the bolt anchorage area, impairs ventilation, impedes the passage of persons, causes injury to miners, causes miners' withdrawal from the area, or disrupts activities for more than one hour.

(i)   All roof control materials should conform with standards as established by various testing agencies such as the Ca-

nadian Standards Association (CSA) or the American Society for Testing and Materials Specifications (ASTMS).
In the absence of standards, such materials could be approved by the regulator.

(j)   The regulator should from time to time issue directions,
        such as found in 30 CFR, respecting the use of roof bolts,
        torquing requirements for roof bolts, and testing require-
        ments for roof bolts and for other types of roof support
        systems.

(k)   All entries and drives where roof bolting is the main means
        of roof support should have imbedded warning devices that
        monitor any downward movement in the roof strata. Such
        warning devices should be of a type approved by the
        regulator and should be placed at intervals specified on the
        plan. Installation of such devices should not relieve the op-
        erator from making regular inspections as prescribed. (The
        type of device referred to here is that generic category in
        which the "tell-tale" extensometer — the simple mechani-
        cal gauge produced at the CANMET Coal Research Labo-
        ratory in Cape Breton — would be included.)

45.   The legislation governing coal mines should be revised to ensure
      that every underground coal mine operator be required to engage,
      when required, the services of a qualified mining engineer with
      specialized postgraduate training in rock mechanics relating to
      coal mines.

46.   The legislation and regulations governing coal mines should be
      reviewed to ensure that all personnel working underground
      receive training in ground control as appropriate to their activities
      and responsibilities. In particular:

(a)   Coal miners should receive a course on ground control as
        part of their basic mine training, plus annual refresher
        courses on ground control.

(b)   Mining supervisory staff, including mine managers, under-
        ground managers, and overmen, should receive extensive
        training in ground control.

(c)   Non-mining personnel employed underground should re-
        ceive sufficient training in ground control to enable them
        to recognize potential hazards.

(d)   Training programs for these three categories of employee
        should be developed by mine management in cooperation
        with the joint occupational health and safety committee

and the regulator. The regulator should review these training programs to ensure that they reflect changing technology and mining practices

47.   The mandate of the Department of Natural Resources should be formally reviewed and clarified vis-à-vis the mandate of the Department of Labour to ensure that there are no gaps in the regulatory process.

48.   A formal procedure should be put in place to provide for adequate communication and cooperation between the Department of Natural Resources and the Department of Labour to ensure that there is adequate provision for all aspects of the regulatory process.

49.   The *Mineral Resources Act* should be amended to identify clearly the role of the Department of Natural Resources in monitoring mine planning in the province. Such a role should encompass the duty to make site inspections to ensure that an operator is mining in conformity with plans approved by the department.

50.   The *Mineral Resources Act* should be amended to identify clearly the role of the Department of Natural Resources in ensuring the "safe" operation of mines in the province.

51.   The province should act to ensure that deputy ministers' positions are adequately described in detailed job descriptions. Such job descriptions should include but not be limited to the following requirements:

    (a)   Upon appointment, the deputy shall forthwith familiarize himself or herself with all the operations of the department as set out in a current organizational chart.

    (b)   The deputy shall have a working knowledge of all the legislation and regulations the department is administering.

    (c)   Where there is more than one department with responsibilities for common projects or interests, the deputy shall ensure that proper procedures are instituted and maintained to provide adequate liaison with the other department or departments, with the result that no gaps exist in the administration of the legislation.

52.   The Department of Natural Resources should no longer act as both promoter and regulator of the development of mineral and energy resources in the province, since this dual mandate constitutes a conflict-of-interest situation. The department should assume the role of helping the developer to formulate a plan that ensures both the safe and the efficient exploitation of the resource.

The department must, first and foremost, work to ensure compliance with the general structure of the legislation in keeping with the purposes for which such legislation was enacted.

53. The structure and staff of the Department of Natural Resources should undergo a complete and intensive review, preferably by an outside agency, with the objective of establishing an efficient and responsible mechanism for the supervision and husbanding of our natural resources.

54. Visits by the inspectorate to the industrial site should not always be subject to prior notice. The inspectorate should schedule visits irregularly, and the operator should expect inspections at any time. Frequency of visits should be dictated by the safety performance of the operator.

55. The unacceptable performance of Claude White and Albert McLean in the conduct of their duties as mine-safety inspectors and regulators, coupled with their demeanour at the Inquiry hearings, must surely have destroyed any confidence the people of Nova Scotia might have had in the department's safety inspectorate. Accordingly, both White and McLean should be removed from any function relating to safety inspection or regulation.

56. The lassitude that paralysed the inspectorate and rendered it ineffectual in dealing with Westray seems deep-seated and pervasive. Therefore, an independent and professional safety consultant should evaluate the inspectorate and its personnel. The consultant should make recommendations for the restructuring of the safety inspectorate and its staff to ensure that the workers and the people of Nova Scotia benefit from a competent, well-trained, and properly motivated safety inspectorate.

57. The *Occupational Health and Safety Act*, 1996, should be revised to incorporate the following changes:

    (a) Except in the case of a demonstrated emergency, any communication respecting health and safety concerns should go initially to the firstline supervisor. If the first-line Supervisor is unable or unwilling to resolve the matter, then the complaint should be taken directly to a member of the joint occupational health and safety committee, for resolution by the committee as expeditiously as possible.

    (b) Provisions should be adopted to clarify how interests of non-union employees in a union shop will be met on the joint occupational health and safety committee.

(c)   No member of management whose principal duty or concern relates to production quotas should be eligible for membership on the joint occupational health and safety committee.

(d)   No member of the executive of any employee organization or union, or any person who has served in such capacity within the preceding year, should be eligible for membership on the joint occupational health and safety committee.

(e)   Provisions should be adopted to define clearly the health and safety obligations of employees to workers on site who are employed by contractors other than the principal employer. Those contractor employees should have obligations similar to those of the employees of the principal employer.

(f)   For greater certainty, the terms "serious injury" and "bodily injury" should be replaced with the one term "serious injury," defined as any injury that requires immediate medical aid or hospitalization or renders the employee unable to perform his or her regular duties for a period in excess of 24 hours.

58.   The province of Nova Scotia should immediately study the British approach to ministerial responsibility, as illustrated by the publication *Questions of Procedure for Ministers* (1992), and move to adopt this type of program. Other jurisdictions should be canvassed for information on similar programs. The program adopted should include a codified and published statement of guidelines for ministers outlining ministerial responsibilities.

(a)   The guidelines for ministers program should be provided to all new ministers. It should include definitions of the nature and extent of the responsibility and accountability for the actions of the department over which a minister presides.

(b)   A minister should have clear guidelines to the frequency and detail of division briefings and the circumstances under which the immediate division head should participate in the briefing along with the deputy minister.

(c)   A minister should have access to independent advice about the nature and the extent of ministerial responsibility in specific situations. Such advice could be provided, ad hoc, by a person with recognized expertise in the field.

59.   Any applicant for an underground coal mining permit should make a clear and unequivocal commitment to the concept of mine safety in the context expressed in the phrase — safe mine production. This clear commitment must be manifest in mine development proposals and plans. Therefore, before a mining permit is granted, the applicant should have to show that it has sufficient financial and other resources to ensure a reasonable margin of safety. The existence of this margin of safety will minimize the possibility that safety measures may be overlooked or avoided to maintain production schedules.

60.   All rules and regulations relating to the operation of coal mines should be contained in Regulations made pursuant to the *Occupational Health and Safety Act*. The *Coal Mines Regulation Act* and the portions of the *Mineral Resources Act* dealing with operations should be repealed.

61.   A legislative review committee should be established to review periodically the underground coal mine regulations to ensure that the regulations reflect current technology and that the use of such technology is consistent with mine safety. The committee should have the power to engage mining consultants with specific expertise consonant with the technical matters being considered. This committee could be modelled after the Mining Legislative Review Committee of the province of Ontario and should contain representation from the provincial departments involved in the planning and regulation of underground coal mines.

62.   The regulator should be given authority to grant exemptions to or variances in the regulations if satisfied that such exemptions or variances will in no way detract from the safety of the miners and other underground workers. The burden is on the mine operator to demonstrate to the satisfaction of the regulator that safety considerations have not been prejudiced.

63.   A mine developer or mine operator should submit all mine plans, including plans for the development, construction, or alteration of an underground coal mine, to the regulator for approval. No such plans should be acted upon or otherwise implemented until they have been approved in writing by the regulator. The regulator may require further detailed plans of the mine or the surrounding geological configurations. The regulator may require that the developer or operator have the plans, or portions of them, reviewed at the expense of the developer by mining consultants having expertise in any or all of the following disciplines: rock

mechanics, mine ventilation, roof control, underground equipment, and electrical applications.

64. The province should take immediate action to reach agreement with the federal Department of Labour for the inspectorate of that department to assume the underground coal mine regulation and inspection functions currently under the aegis of the provincial Department of Labour.

65. The province should collaborate with the federal Department of Labour to draft updated underground coal mining regulations applicable to all coal mines in Nova Scotia. These common regulations would then be administered throughout the province by the inspectorate at present functioning under the provisions of the *Canada Labour Code* regulations. Such regulations should be drafted with the advice and assistance of competent coal mining professionals with demonstrated expertise in the various fields of ventilation, ground control, electrical applications, training, and mine rescue.

66. If it is decided to pattern the Nova Scotia coal mine regulation regime after that of the United Kingdom, all mine inspectors should have at least a degree in mining engineering, with some specialist training in both rock mechanics and ventilation relating to underground coal mining. If the U.S. Mine Safety and Heath Administration approach is adopted, all mine inspectors should receive adequate initial training. In either case, all mine inspectors should be required to take periodic training, of at least one week per year, at an institute specializing in mine inspection and safety.

67. Labour and management should work together to educate and regulate the underground miner with a view to eradicating the practice of smoking in the coal mining environment. The following requirements should apply:

    (a) Tobacco smoking and the possession of smoking materials and lighters by any person underground should be grounds for immediate dismissal from employment, the reason for dismissal to be recorded in the employee's record.

    (b) Proof of tobacco smoking underground or possession of smoking materials underground should provide sufficient grounds for dismissing any grievance taken by an employee for unjust dismissal, and any arbitrator should be prohibited from substituting any other penalty in lieu of dismissal.

      (c)   Labour and management, with the cooperation of the Department of Labour, should investigate the feasibility of acquiring tobacco detection devices that would monitor miners entering the mine.

68.   Every mine operator, indeed, every industrial plant or facility, should have a well-defined and comprehensive emergency procedures manual containing a complete and up-to-date list of all persons involved in any emergency operation. This list should contain an organization chart listing the individuals and their respective tasks, and a current telephone listing for each person. The manual should be prepared by the company in consultation with both the joint occupational health and safety committee and the safety coordinators with the appropriate government departments. The manual should set out, in detail, the quantity and location of all emergency supplies and equipment and the details of the deployment of these materials. A current copy of any such approved emergency procedures manual should be filed with the director of occupational health and safety, and copies should be provided to each person assigned any duty under the manual.

69.   The Department of Labour, in consultation with the operator, should establish such rules and regulations that would ensure the department a full and active role in every mine-related emergency procedure or rescue operation in the province. The rules and regulations should set out the duties and responsibilities of each department inspector or safety examiner in any mine-related emergency or rescue operation.

70.   Rescue and emergency equipment should be standardized so that those persons trained in rescue procedures will be completely familiar with the equipment available. Similarly, the various testing devices should be standardized so that the rescuers are able to use these devices without losing valuable time and without the danger of mistaken or inaccurate readings.

71.   Every community at or near which underground mining operations are carried out should have a plan to provide emergency medical, fire, and other support services. The plan should include providing emergency training to the appropriate people in those communities. Some familiarity with the underground environment could be helpful in the event of a disaster.

72. Mine-rescue competitions, long a fixture in the underground mining industry, provide a valuable training incentive for miners. These competitions should be continued.

73. The Government of Canada, through the Department of Justice, should institute a study of the accountability of corporate executives and directors for the wrongful or negligent acts of the corporation and should introduce in the Parliament of Canada such amendments to legislation as are necessary to ensure that corporate executives and directors are held properly accountable for workplace safety.

74. The province of Nova Scotia should review its occupational health and safety legislation and take whatever steps necessary to ensure that officers and directors of corporations doing business in this province are held properly accountable for the failure of the corporation to secure and maintain a safe workplace.

INDEX

B

BACKGROUND TO BILL C-45
- legal response to Westray mine disaster, 18-24
- • charges under *Occupational Health and Safety Act* stayed or dropped, 20-21
- • chronology of charges, 18, 155-165
- • criminal charges against corporation and managers, 21
- • • charges stayed for lack of particularity, 21
- • • new charges proceeding to trial, 21-23
- • investigations and prosecutions, 18, 21
- • Public Inquiry, 19-20, 24-26
- • • constitutionality challenged on basis of potential criminal charges, 19
- • • injunction against commencement of proceeding, 19
- • • investigation into causes of mine explosion, and whether preventable, 24
- • • recommendations, 25-26, 167-183
- • • rhetorical "what if" questions, 25
- • • Supreme Court of Canada permitting Inquiry to proceed, 19-20
- • • union participating on behalf of workers, 24
- • • union urging creation of new criminal offence, 24-25
- legislative responses, 26-37
- • Bill C-45, introduction of, 33-37
- • • criminal law reserved for most serious offences, 35
- • • highlights of Bill C-45, 34
- • • "identification theory" limiting responsibility to most senior management, 35
- • • individuals engaging corporate responsibility extended, category of 36
- • • legislative development and history, 36-37
- • • legislative summary, 34n
- • • NDP Private Members bills, 33
- • • OHS criminal negligence, new offence of, 35
- • • OHS legal duty, when breached, leading to charge of OHS criminal negligence, 36
- • • OHS regulatory laws, continuing reliance on, 35
- • • "organization" term replacing "corporation" term, 35
- • • "representative" term extending categories of culpable individuals, 36

- • • "senior officer", having knowledge of commission of offence, 34
- • • "senior officer" not preventing organization from committing offence, 35
- • Bill C-259, 28
- • • corporation becoming liable for criminal offence, scenarios in which, 28-29
- • • directors and officers of corporation, new offences for, 29
- • • maximum penalty, 29
- • • OHS criminal liability for corporations, directors and officers, 28
- • Bill C-284, 29-30, 59-60
- • • corporate culture model of liability, 59-60
- • • criminal liability for management of corporation, 30
- • • directing mind theory of corporate liability, moving away from, 29-30
- • • reverse onus raising constitutional and *Charter* concerns, 30
- • • staff of corporation committing offence resulting in corporate liability, 29, 30
- • new Nova Scotia *Occupational Health and Safety Act*, 26-27
- • • directors and corporate executives, no legal duties for, 27
- • • health and safety committees' role enhanced, 27
- • • internal responsibility system defined, 27, 103
- • • new legal duties and precautions, 27
- • Private Member's Motion 455, 27-28
- • • request to amend *Criminal Code* in accordance with Recommendation 73, 27-28
- • Standing Committee on Justice and Human Rights, 30-33
- • • corporate criminal liability examined, 30-31
- • • internal responsibility system and its enforcement criticized, 31, 103
- • • principles guiding legislation on corporate criminal liability, 31
- • • recommendation that corporate criminal liability be reformed, 32
- • • Response Paper to Committee's Report, 32-33
- • • review of legislation respecting criminal justice and human rights, 30
- • • workplace intimidation preventing workers from enforcing their rights, 31
- • Westray mine disaster, 15-18
- • • Curragh developing mine, 15-16

• • • business case not as strong if higher
standards of worker safety, 16
• • • hazards, prevalence of, 16
• • • production figures too optimistic, 16
• • • work not stopping, 16-17
• • danger and death in coal mining region,
17
• • • alternative being unemployment, 17
• • • challenge of finding sustained full-time
work, 17
• • generally, vii, 1-2, 12
• • methane gas explosion, 17-18
• • • coal dust causing explosion to roll
through mine, 18
• • • gas highly flammable, 17
• • • government inspectors requesting
corrective action, 18
• • • proper ventilation required to keep gas
at manageable level, 18
• • miners killed by explosion, fire and
ground collapse, 15
• • power to close mine with government
OHS regulators, 18
• • proper safety mining procedures not
followed, 15

BILL C-45: DETAILED ANALYSIS *See
also* **BACKGROUND TO BILL C-45**
• accessoryship provision repealed, 75
• coming into force date, 94-95
• definitions, 63-65
• • "every one", 63, 64
• • "organization", 35, 63-65
• • "representative", 36, 64, 65
• • "senior officer", 34-35, 64, 65
• extending potential criminal liability,
64-65
• • liability extending beyond corporations
to any organization, 64-65
• • non-profit organizations, no exemption
for, 65
• • specific types of bodies included, 65
• false pretences, 74-75
• • "corporations" word replaced by
"organization" term, 74-75
• introduction, 63
• OHS legal duty, 68-73
• • aiding, abetting, counselling, attempting
or being accessory to offence, 71
• • authority to direct, 68-69, 72
• • • organizational hierarchy to be
examined to determine reporting
authority, 69
• • • particular title or level of responsibility
not specified, 69
• • • "senior officer" term not used, resulting in broader category, 69
• • breach of duty giving rise to OHS criminal negligence, 69, 71

• • broad and overarching legal duty, 71-72,
100
• • • extending from most senior decision-
makers to foreman and co-workers,
71-72
• • causal connection, 70
• • components not defined, 72, 99-100
• • • "any other person", 72
• • • "has authority", 72, 99-100
• • • "reasonable steps", 77, 99
• • directing "how" another person doing
work or performing task, 68
• • • "how" synonymous with "by what
means", 68
• • due diligence, defence of, 70
• • "every one", duty applying to, 68
• • harm, level of, 41, 72-73
• • • "bodily harm" defined, 41, 73
• • no new specific steps, actions, control
requirements or OHS systems added,
70-71
• • • reinforcing existing OHS statutes and
regulations, 71
• • public and workers, 70, 72, 99, 100
• • "reasonable steps" to prevent bodily
harm, taking, 69, 72
• • • compliance with OHS regulations and
industry standards, 69
• • • not defined, 72
• • "that person", scope of phrase, 69-70, 72
• • • "any other person", 70, 72
• • trespasser not covered, 70
• • work area, scope of, 73
• organizational guilt for offence where fault
element not negligence, 66, 67-68
• • fault element implied or specified, 67
• • senior officer intending to benefit corporation, 67-68
• • three alternative means by which organization may be found guilty, 68
• • traditional criminal charge, 67
• organizational guilt for offence with fault
element of negligence, 65-67
• • identification theory applying prior to
Bill C-45, 66
• • representative being party to offence,
and senior officer departing markedly
from standard, 66-67
• • section 22.1 applying to OHS legal duty
and OHS criminal negligence, 66
• • standard for corporate criminal liability,
67
• • • marked departure from standard of
care, 67
• preliminary inquiry, procedure on, 76-77
• • "corporations" word replaced by
"organization" term, 76-77
• proceeds of crime and persons deemed
absconded, 76

• • "corporations" word replaced by
"organization" term, 76
• procuring attendance, 80-81
• • service of process, 80-81
• • • "corporations" word replaced by
"organization" term, 80-81
• selling defective stores to Her Majesty,
75-76
• • "corporations" word replaced by
"organization" term, 75-76
• sentencing generally, 85-93
• • discharges, absolute and conditional, 86-
87
• • • "corporations" word replaced by
"organization" term, 86-87
• • fines, increasing limit of possible, 92-93
• • • "corporations" word replaced by
"organization" term, 92-93
• • • enforcement of unpaid fines, 93
• • • factors in determining appropriate fine,
92
• • • maximum fines increased, 93
• • fines, powers to impose, 91-92
• • • "corporations" word replaced by
"organization" term, 91-92
• • previous convictions, evidence of, 86
• • • "corporations" word replaced by
"organization" term, 86
• • probation, "optional conditions" of,
87-91
• • • compliance order with reasonable
conditions imposed, 91
• • • "optional conditions" definition
extended, 87-88
• • • optional terms for organizations convicted of criminal offences, 88-89
• • • policies, standards and procedures
imposed to reduce likelihood of
offence, 89-90
• • • post-conviction scrutiny by courts and
OHS regulators, higher level of,
88-89
• • • public shame orders, 90
• • • reporting to court on implementation
of policies, 89, 90
• • • restitution order, 89
• • • supervision of development and
implementation of OHS policies, 91
• • probation reports, 85-86
• • • "corporations" word replaced by
"organization" term, 85-86
• sentencing of organization, factors to
consider in, 81-85
• • criteria for sentencing organizations,
82-83
• • • mandatory considerations, 82-83
• • judicial discretion, 82
• • OHS prosecutions, factors in sentencing
in, 83
• • proportionality, principle of, 82

• • sentencing factors under Bill C-45, 83-
85, 142
• • • advantage realized by organization,
83-84
• • • converting or concealing assets, 84
• • • cost of prosecution and investigation,
84
• • • economic impact of sentence, 84
• • • measures to reduce likelihood of
offences, 85
• • • penalty imposed by organization on
representative, 84-85
• • • planning, degree of, 84
• • • previous convictions or regulatory
sanctions, 84
• • • regulatory penalty, 84
• • • restitution order, 85
• • text of s. 718.21, 81
• summary conviction offences, 93-94
• • appearances, 93-94
• • • "corporations" word replaced by
"organization" term, 93-94
• text of provisions, 145-154
• theft, 73-74
• "corporations" word replaced by "organization" term, 73-74
• trial with jury, 79-80
• • appearances and notice, 79-80
• • • "corporations" word replaced by
"organization" term, 79-80
• • presence in court during trial, 80
• • • "corporations" word replaced by "organization" term, 80
• trial without jury, 77-78, 94
• • appearances and elections, 77-78, 94
• • • coordinating amendment, 94
• • • "corporations" word replaced by
"organization" term, 77-78, 94
• warrants of committal, 78
• • "corporations" word replaced by "organization" term, 78

BODILY HARM
• defined, 41, 73
• interference with comfort, element of, 41
• mental and emotional faculties, 41
• occupational disease, 73
• ordinary meaning, 41
• psychological injury, 41
• transient or trifling, more than, 41, 73

C

COUNSEL, RIGHT TO *See also*
ENFORCEMENT PROCEEDINGS
UNDER *CRIMINAL CODE*
• *Charter*, s. 10(b), 130
• fundamental right, 130
• legal aid or duty counsel, informing of
availability of, 131

• right to consult with lawyer, 130-131
• right to remain silent, 131

CRIMINAL NEGLIGENCE, LAW OF
• acts and omissions, 42-43
• • distinction, potential for, 42
• • fault element for both being same, 42-43
• • • contributing to death, 43
• • • no distinction between negligent conduct and failure to act, 43
• • • uniform objective standard, 42-43
• corporate criminal liability, 49-51, 56-61
• • basis of corporate criminal liability in Canada, 50
• • "corporation" word replaced by "organization" term, 51
• • directing mind of corporation, acts authorized by, 50
• • identification theory, 50
• • • corporations responsible for acts of senior officers or directing minds, 50
• • lack of clear direction in *Criminal Code*, 49-50
• • models of corporate liability, 56-61
• • • corporate culture model, 59-60
• • • corporate killing model, 60-61
• • • identification theory, 50, 51-52, 56-57
• • • vicarious liability model, 58-59
• • trend toward broader concept of corporate responsibility, 50-51
• fault element for organizations and corporations, 51-55
• • fault element implied or specified, 51
• • identification theory of organizational criminal liability, 51-52
• • • corporation not liable if senior officer acting against corporation's interests, 52
• • • fault element of senior officer attributed to organization, 51-52
• • regulation of organizations in different ways, 51
• • standard of corporate liability for criminal negligence, 55
• • • although breach of duty, evidence failing to prove fault element, 55
• • who in corporation having fault element so that corporation having fault element, 52-54
• • • criminal negligence involving workplace accident, 52-54
• • • directing mind test, 52, 54
• objective vs. subjective standard, 44-47
• • objective standard prevailing, 47
• • objective, subjective or both, 44
• • Ontario Court of Appeal, 46-47
• • • objective standard adopted, 46-47
• • Supreme Court of Canada divided, 44-46
• • • marked departure from norm or standard, 44, 45

• • • minimum intent requirement of awareness or advertence, 45-46
• offence of criminal negligence generally, 39-42, 119-120
• • acts or omissions of accused, 39, 40, 42
• • • no distinction between acts or omissions, 42
• • "bodily harm" defined, 41
• • • interference with comfort, element of, 41
• • • mental and emotional faculties, 41
• • • ordinary meaning, 41
• • • psychological injury, 41
• • • transient or trifling, more than, 41
• • *Criminal Code*, ss. 219-221, 40
• • example of ergonomically incorrect work station causing back problems, 41-42
• • • whether back strain constituting bodily harm, 42
• • example of employer failing to prohibit outright smoking, 42
• • • whether respiratory diseases constituting bodily harm, 42
• • fault element of offence being problematic, 42
• • • objective, subjective or both, 42
• • legal duties imposed under *Criminal Code*, 39-40
• • • breach of duty forming basis of criminal negligence, 39
• • OHS legal duty and OHS criminal negligence, 39
• • wanton or reckless disregard for lives or safety of others, 39, 40
• • • motor vehicle accidents, 40
• wanton or reckless disregard, meaning of, 47-49
• • corporation, context of, 49
• • differing views, 48-49
• • expansive approach, 47-48
• • "reckless" defined, 48, 49
• • synthesizing phrase "wanton or reckless disregard", 48
• • "wanton" defined, 48, 49

D

DUE DILIGENCE DEFENCE
• Bill C-45 "reasonable steps" and OHS due diligence, 118-120
• • compliance with new duty requiring complying with OHS law, 119
• • • interpretation of reasonable steps problematic with no national consistent statute, 119
• • • OHS legislation varying among jurisdictions, 119
• • criminal negligence, elements of, 119-120

• • criminal standard to be breached and proven for conviction, 119, 120
• • • marked departure from norm, 120
• • fact-specific determination, 119
• • failure to take reasonable steps, 118, 119
• • • failure to comply with OHS law not necessarily resulting in criminal conviction, 119
• • no reverse onus for OHS criminal negligence, 118
• generally, 70
• onus of proof on defendant to establish defence of due diligence, 110, 113
• reverse onus, validity of, 113-115
• • Crown proving *prima facie* case, and then onus of proof shifting, 114-115
• • • civil standard of proof on defendant, 114, 115
• • whether violation of *Charter*, 113-114, 115
• • *Wholesale Travel Group* case, 113-114
• • • reverse onus on strict liability offence not offending *Charter*, 114, 115
• strict liability offences, 109, 110
• two separate branches of defence, 110, 111-112
• • first branch: mistaken fact branch, 110, 111-112
• • • *Ontario v. London Excavators & Trucking Ltd.* case, 111-112
• • second branch: reasonable precautions branch, 110, 112-113, 118
• • • assessment of whether company acting reasonably, 112-113
• • • compliance with internal responsibility system, 112

DUTY OF PERSONS DIRECTING WORK
• aiding, abetting, counselling, attempting or being accessory to offence, 71
• authority to direct, 68-69, 72
• • organizational hierarchy to be examined to determine reporting authority, 69
• • particular title or level of responsibility not specified, 69
• • "senior officer" term not used, resulting in broader category, 69
• breach of duty giving rise to OHS criminal negligence, 69, 71
• broad and overarching legal duty, 71-72, 100
• • extending from most senior decision-makers to foreman and co-workers, 71-72
• causal connection, 70
• components not defined, 72, 99-100
• • "any other person", 72
• • "has authority", 72, 99-100
• • "reasonable steps", 77, 99

• directing "how" another person doing work or performing task, 68
• • "how" synonymous with "by what means", 68
• due diligence, defence of, 70
• "every one", duty applying to, 68
• generally, 104
• harm, level of, 41, 72-73
• • "bodily harm" defined, 41, 73
• • • occupational disease, 73
• no new specific steps, actions, control requirements or OHS systems added, 70-71
• • reinforcing existing OHS statutes and regulations, 71
• OHS legal duty, when breached, leading to charge of OHS criminal negligence, 36
• public and workers, 70, 72, 99, 100
• "reasonable steps" to prevent bodily harm, taking, 69, 72
• • compliance with OHS regulations and industry standards, 69
• • not defined, 72
• "that person", scope of phrase, 69-70, 72
• • "any other person", 70, 72
• • • public and workers, 70, 72
• trespasser not covered, 70
• work area, scope of, 73

E

ENFORCEMENT PROCEEDINGS UNDER *CRIMINAL CODE*
• arbitrary detention, right to be free from, 127
• • right to counsel, 127
• arrest of accused, 128-130
• • common law power of arrest, 129
• • compelling attendance at court, 129-130
• • corporations, service on, 130
• • judicial warrant for arrest, 128-129
• • police officer's powers of arrest, 128
• • photographing and fingerprinting, 129
• • warrant to enter dwelling-house, 129
• • • judicial authorization based on reasonable grounds, 129
• • • response to *R. v. Feeney* decision, 129
• • • warrantless arrest in dwelling-houses prohibited under *Charter*, 129
• • • privacy interest in dwelling outweighing interests of police, 129
• bail: judicial interim release, 131-133
• • offences listed under s. 469 of *Code*, 131, 132
• • • application for release to judge required, 132
• • • no automatic release hearing, 132
• • offences not listed under s. 469 of *Code*, 131-132

• • • automatic right to bail hearing, 131-132
• • pre-trial detention given full credit in sentencing, 133
• • pre-trial release status, determining, 131
• classification of offences, 123-125
• • dual or hybrid offences, 123
• • • Crown's right to select, 123
• • indictable offences, 123, 124-125
• • • three categories, 124
• • summary conviction offences, 123-124
• conclusions, 143-144
• • OHS protection given greater attention and importance, 143
• • seriousness of being charged balanced by *Charter* rights, 143
• • workers having healthy and safe workplace, ensuring, 143
• criminal negligence, accused's election in, 134-136
• • judge alone, 135
• • jury trial, 135
• • preliminary inquiry, where, 135
• • provincial court only, 135
• • provincial court or superior court, 134-135
• • re-election by accused, 136
• • superior court only, 135
• disclosure by Crown, 133-134
• • accused's right to make full answer and defence, 133
• • delay or denial of production, 134
• • Crown obligation to disclose all relevant information, 133-134
• • • early disclosure and ongoing disclosure, 133
• • • logically probative information, 133
• • • power, possession or control, information in its, 133-134
• • stay of proceedings if failure to disclose, 133
• • third party disclosure, 134
• • • application where evidence in possession or control of, 134
• initiation of criminal proceedings, 127-128
• • indictment, issuing of, 127, 128
• • information, swearing of, 127, 128
• • • summary conviction offences, 128
• • lawyer retained by accused, 127-128
• introduction to criminal justice system, 121-123, 124-125
• • *Criminal Code* setting standards of criminal behaviour across Canada, 123
• • criminal investigation process, 124-125
• • federal power over criminal law and procedure, 121-122
• • federal works and undertakings, 122
• • judges, powers of, 122
• • • provincial court judges, 122
• • • superior court judges, 122

• • policing, 122
• • • national police force, 122
• • • provincial police forces, 122
• • prosecution of offences, 122
• • • federal attorney general prosecuting drug offences, 122
• • • provincial attorney general prosecuting *Criminal Code* offences, 122
• • provinces not permitted to enact laws having fault element, 122
• preliminary inquiry, 135, 136-137
• • election of accused, 136
• • mandatory, where, 136-137
• • opportunity for accused to learn about Crown's case, 137
• • opportunity for Crown to discover theory of defence, 137
• • organization appearing by counsel or agent, 137
• • sufficiency of evidence to convict, 137
• pre-trial conference, 137-138
• • used to identify and narrow issues, 138
• right to counsel, 130-131
• • *Charter*, s. 10(b), 130
• • fundamental right, 130
• • legal aid or duty counsel, informing of availability of, 131
• • right to consult with lawyer, 130-131
• • right to remain silent, 131
• sentencing, 141-143
• • discharge, 142
• • factors to consider in sentencing of organization, 81-85, 142
• • fines, 142
• • OHS policies, standards and procedures imposed, 143
• • prescribed penalty, 142
• • public disclosure, 143
• • restitution, 142
• • restorative justice, 141-142
• trial, 138-141
• • adversarial system, 138
• • arraignment of accused, 139
• • closing statements, 141
• • Crown's case, 140
• • • examination-in-chief, cross-examination and re-examination, 140
• • Crown's opening statement, 139-140
• • • outline of case, 139-140
• • defence's case, 140-141
• • defence's opening statement, 140
• • guilty plea, consequences of, 139
• • jury trial, 139, 141
• • • instructions to jury, 141
• • rebuttal evidence by Crown, 141
• • right to trial within reasonable time, 138-139
• • • *Charter*, s. 11(b), 138-139
• • • prejudice required where unreasonable delay, 138-139

- unreasonable search and seizure, protection against, 125-127
- • *Charter*, s. 8, 125
- • DNA warrant required for taking bodily substances, 125
- • • reasonable grounds to believe that designated offence committed, 125
- • exclusion of evidence acquired in illegal search, 126
- • • whether admission of evidence bringing administration of justice into disrepute, 126
- whether admission of evidence undermining accused's right to fair trial, 126-127
- • private property, right to enter, 126
- • • owner's or occupier's consent, 126
- • • search warrant, valid, 126
- • reasonable expectation of privacy, 125

F

FAULT ELEMENT FOR ORGANIZATIONS AND CORPORATIONS, 42-43, 51-55 *See also* **CRIMINAL NEGLIGENCE, LAW OF**

I

IDENTIFICATION THEORY *See also* **CRIMINAL NEGLIGENCE, LAW OF**
- criticisms of narrow and simplistic nature, 36
- fault element for organizations and corporations, 51-52
- • corporation not liable if senior officer acting against corporation's interests, 52
- • fault element of senior officer attributed to organization, 51-52
- • generally, 50, 51-52
- • organizational guilt for offence with fault element of negligence, 66
- • responsibility limited to most senior management, 35
- senior officers or directing minds, acts or omissions of, 50, 52, 54, 56, 57
- • executive level authority, 56
- • board of directors, managing director, manager, 57
- • manager of used car lot, 57

INTERNAL RESPONSIBILITY SYSTEM *See also* **OCCUPATIONAL HEALTH AND SAFETY IN CANADA**
- "direct responsibility", 101
- "contribution system", 101
- core elements, 101

- • legal duties on various workplace stakeholders, 101
- criticism at Westray Inquiry, 31, 103
- definition in Nova Scotia OHS statute, 27, 103
- foundational concept of OHS law, 104
- generally, 98
- health and safety committees and representatives, 102
- • inspection of workplace and investigation of accidents, 102
- toxic substances and controlled products regulated, 102-103
- • WHMIS, 102-103
- • • national consistent program, 102-103
- • • training, labelling, and providing material safety data sheets, 102
- workplace stakeholders having responsibility, 101

N

NEGLIGENCE *See* **CRIMINAL NEGLIGENCE, LAW OF**

O

OCCUPATIONAL HEALTH AND SAFETY LAW IN CANADA
- enforcement of OHS law and due diligence, 108-115
- • Bill C-45 vs. OHS law, 108
- • due diligence defence, 109-115
- • • Bill C-45 "reasonable steps" and OHS due diligence, 118-120
- • • onus of proof on defendant to establish defence of due diligence, 110, 113
- • • reverse onus, validity of, 113-115
- • • strict liability offences, 109, 110
- • • two separate branches of defence, 110, 111-112
- • government regulators issuing orders or directions, 108
- • quasi-criminal prosecution of OHS regulatory offences, 108-109
- • • strict liability offences or public welfare offences, 108-109
- • sentencing and maximum fines, 115-118
- • • federal, 115
- • • Northwest Territories, 118
- • • Nunavut Territory, 118
- • • provincial jurisdictions, 115-117
- • • Yukon Territory, 117-118
- • types of offences and defence of due diligence, 109-110
- • • absolute liability offences, 109-110
- • • *mens rea* offences, 109
- • • strict liability offences, 109, 110-111
- general duty clauses in OHS statutes, 104-108

• • federal jurisdiction, 105
• • • general duty to employees, 105
• • Northwest Territories and Nunavut, 107
• • provincial jurisdictions, 105
• • • Alberta, 105
• • • British Columbia, 105
• • • Manitoba, 106
• • • New Brunswick, 106
• • • Newfoundland, 107
• • • Nova Scotia, 107
• • • Ontario, 106
• • • Prince Edward Island, 106-107
• • • Quebec, 106
• • • Saskatchewan, 105
• • • Yukon Territory, 107
• • similarities among various Canadian jurisdictions, 104
• • similarities to s. 217.1 of *Criminal Code*, 104, 107-108
• • statement requiring employers to take reasonable precautions, 104
• internal responsibility system, 98, 101-104
• • "direct responsibility", 101
• • "contribution system", 101
• • core elements, 101
• • • legal duties on various workplace stakeholders, 101
• • criticism at Westray Inquiry, 103
• • definition in Nova Scotia OHS statute, 103
• • foundational concept of OHS law, 104
• • health and safety committees and representatives, 102
• • • inspection of workplace and investigation of accidents, 102
• • toxic substances and controlled products regulated, 102-103
• • • WHMIS, 102-103
• • workplace stakeholders having responsibility, 101
• introduction, 97-100
• • *Constitution Act, 1867* having no specific jurisdictional designation, 97-98
• • • 10 per cent of workplaces being federally regulated, 97
• • • 90 per cent of workplaces being provincially regulated, 97-98
• • duty on workplace stakeholders and employers to provide safe workplace, 99-100
• • • *Criminal Code*, 99-100
• • • employees vs. workers, 99
• • • workers' compensation legislation terminating worker's right to sue for breach, 99
• • external responsibility system, 98
• • • means of enforcement, 98
• • • stakeholders in workplace, 98-99
• • internal responsibility system, 98

• • workers' compensation for injured workers, 97, 98

OHS LEGAL DUTY *See* **DUTY OF PERSONS DIRECTING WORK** and **OCCUPATIONAL HEALTH AND SAFETY LAW IN CANADA**, general duty clauses

OVERVIEW *See also* **PURPOSES OF BILL C-45**
• Bill C-45, vii, viii
• • corporate criminal liability, vii-viii
• • • "identification theory" replaced, vii, 12
• • • negligence and subjective fault element, vii
• • "corporation" term replaced by "organization" term, 2
• • criminal investigation and criminal prosecution process, viii
• • criminal liability added, 2
• • legal duty to take "reasonable steps" to protect workers, vii, viii, 1, 2
• • • Canadian workplaces, in all, vii
• • maximum penalties, 1
• • occupational health and safety criminal negligence, new crime of, vii, viii, 1
• • • defence of due diligence not available, viii
• • "organization" defined, 1
• • Royal Assent and proclamation, 1
• • section-by-section analysis, viii
• Bill C-46 and corporate crime, 5-7
• • anti-reprisal provision preventing retaliatory action, 6
• • confidential or privileged information produced pursuant to *ex parte* order, 6-7
• • • CGA-Canada concerns, 6-7
• • court ordered production of documents or data, 6
• • financial sector fraud, organizational criminal liability for, 5
• • insider trading, new criminal offence of, 5-6
• • lack of public and political support, 7
• Canadian criminal law, 2-5
• • *Charter of Rights and Freedoms*, 3
• • *Criminal Code*, 2-3
• • criminal offences, 3-4
• • • *actus reus*, 3, 4
• • • *mens rea*, 3-4
• • defences, 4
• • • excuse or justification, 4
• • • mistake of fact, 4
• • plea of guilty or not guilty, 4
• • regulatory offences or public welfare offences, 3
• • regulatory statutes and offences promoting public good, 3
• • sentencing, 4-5

• • • judge's discretion, 4
• • • maximum penalty, 4
• • • minimum penalty, 4-5
• • • punishment proportionate to seriousness of offence, 5
• • trial process, 4
• • • defence of accused, 4
• • • finding of guilty or not guilty, 4
• • • prosecutor's case, 4
• criminal justice system, 2
• occupational health and safety law in
Canada, viii, 2
• • internal and external responsibility
systems, viii, 2
• Westray Bill, vii, 1
• Westray mine disaster, vii, 1-2

P

PUBLIC INQUIRY
• constitutionality challenged on basis of
potential criminal charges, 19
• injunction against commencement of
proceeding, 19
• investigation into causes of mine explosion, and whether preventable, 24
• recommendations, 25-26, 167-183
• • accountability of corporate executives
and directors, 26, 183
• • criminal liability of corporate executives
and directors, 25
• • health and safety regulations, 26
• • mining plans and permits, 26
• • monitoring and controlling methane gas
levels and coal dust, 25
• rhetorical "what if" questions, 25

PURPOSES OF BILL C-45
• interpretation of Bill C-45 generally, 7-8
• Plain Language Guide, 8, 10-11, 13
• • criminal liability of organizations, 11
• • example of how organization being held
accountable, 11
• political statements and legislative debates, use of, 7-8, 9-10
• • little evidentiary weight but relevant as
to purpose of legislation, 8
• • second reading debate, 9-10
• • • consensus building process, 9-10
• • • criminal proceedings, fruitless, 9
• • • new OHS duty to prevent bodily harm
in workplace, 10
• press release, 9
• summary, legislative, 8-9, 11-13
• • four purposes of Bill C-45, 9
• • "identification theory" of corporate
criminal liability criticized, 12

• • new health and safety crime, 13
• • •OHS criminal negligence, 13
• • new OHS legal duty, 13
• • union advocating enhanced criminal
accountability of directors and officers,
12
• • Westray mining disaster receiving brief
reference, 12

S

SENTENCING *See also* **BILL C-45:
DETAILED ANALYSIS**
• *Criminal Code*, 81-93, 141-143
• • factors to consider in sentencing of
organization, 81-85, 142
• • generally, 141-143
• • probation, "optional conditions" of,
87-91
• OHS laws, 115-118

W

WESTRAY MINE DISASTER *See also*
**BACKGROUND TO
BILL C-45**
• Curragh developing mine, 15-16
• • business case not as strong if higher
standards of worker safety, 16
• • hazards, prevalence of, 16
• • production figures too optimistic, 16
• • work not stopping, 16-17
• danger and death in coal mining region, 17
• • alternative being unemployment, 17
• • challenge of finding sustained full-time
work, 17
• generally, vii, 1-2, 12
• legal response to Westray mine disaster,
18-24
• methane gas explosion, 17-18
• • coal dust causing explosion to roll
through mine, 18
• • gas highly flammable, 17
• • government inspectors requesting
corrective action, 18
• • proper ventilation required to keep gas at
manageable level, 18
• miners killed by explosion, fire and
ground collapse, 15
• power to close mine with government
OHS regulators, 18
• proper safety mining procedures not
followed, 15

**WORKPLACE HAZARDOUS
MATERIAL INFORMATION
SYSTEM (WHMIS),** 102-103